CAMBRIDGE LIBRARY COLLECTION

Books of enduring scholarly value

Earth Sciences

In the nineteenth century, geology emerged as a distinct academic discipline. It pointed the way towards the theory of evolution, as scientists including Gideon Mantell, Adam Sedgwick, Charles Lyell and Roderick Murchison began to use the evidence of minerals, rock formations and fossils to demonstrate that the earth was older by millions of years than the conventional, Bible-based wisdom had supposed. They argued convincingly that the climate, flora and fauna of the distant past could be deduced from geological evidence. Volcanic activity, the formation of mountains, and the action of glaciers and rivers, tides and ocean currents also became better understood. This series includes landmark publications by pioneers of the modern earth sciences, who advanced the scientific understanding of our planet and the processes by which it is constantly re-shaped.

A Sketch of the Geology of Cornwall

Cornwall has one of the oldest mining histories in Europe. At one time, the county was a leading producer of tin, with over 2,000 mines in operation, but competition from overseas saw the boom years of the mid-nineteenth century give way to steady decline. Brenton Symons (1832–c.1908), an experienced mining engineer and metallurgist, firmly believed that the mineral wealth of Cornwall was far from exhausted and that careful application of financial investment and skilled personnel could boost the county's prosperity. This illustrated monograph, published in 1884, is his account of Cornwall's geological characteristics, giving details of the formation, location and economic use of various mineral deposits, and describing the extraction techniques of the key mining districts. Accessible and authoritative, this book remains relevant to readers interested in mineralogy, mining and the economic history of Cornwall.

Cambridge University Press has long been a pioneer in the reissuing of out-of-print titles from its own backlist, producing digital reprints of books that are still sought after by scholars and students but could not be reprinted economically using traditional technology. The Cambridge Library Collection extends this activity to a wider range of books which are still of importance to researchers and professionals, either for the source material they contain, or as landmarks in the history of their academic discipline.

Drawing from the world-renowned collections in the Cambridge University Library and other partner libraries, and guided by the advice of experts in each subject area, Cambridge University Press is using state-of-the-art scanning machines in its own Printing House to capture the content of each book selected for inclusion. The files are processed to give a consistently clear, crisp image, and the books finished to the high quality standard for which the Press is recognised around the world. The latest print-on-demand technology ensures that the books will remain available indefinitely, and that orders for single or multiple copies can quickly be supplied.

The Cambridge Library Collection brings back to life books of enduring scholarly value (including out-of-copyright works originally issued by other publishers) across a wide range of disciplines in the humanities and social sciences and in science and technology.

A Sketch of the Geology of Cornwall

Including a Brief Description of the Mining Districts,
and the Ores Produced in Them

BRENTON SYMONS

CAMBRIDGE
UNIVERSITY PRESS

CAMBRIDGE
UNIVERSITY PRESS

University Printing House, Cambridge, CB2 8BS, United Kingdom

Published in the United States of America by Cambridge University Press, New York

Cambridge University Press is part of the University of Cambridge.
It furthers the University's mission by disseminating knowledge in the pursuit of
education, learning and research at the highest international levels of excellence.

www.cambridge.org
Information on this title: www.cambridge.org/9781108066808

This edition first published 1884
This digitally printed version 2013

ISBN 978-1-108-06680-8 Paperback

The original edition of this book contains a number of colour plates,
which have been reproduced in black and white. Colour versions of these
images can be found online at www.cambridge.org/9781108066808

GEOLOGY OF CORNWALL.

A SKETCH

OF THE

GEOLOGY OF CORNWALL,

INCLUDING A BRIEF

DESCRIPTION OF THE MINING DISTRICTS,

AND THE ORES PRODUCED IN THEM.

BY

BRENTON SYMONS, F.C.S., Assoc. Mem. Inst. C.E.,

MINING ENGINEER AND METALLURGIST.

*Author of "Caradon Mines," "Mining in the East,"
"Hydro-Metallurgical Processes," "Campiglia Mines," &c.*

With Geological Map of Cornwall,

AND

NUMEROUS STEEL PLATES, ILLUSTRATIVE OF INFLUENCE
OF ROCK FORMATIONS ON SCENERY.

———

REPRINTED FROM THE "*GAZETTEER OF CORNWALL.*"

———

LONDON :

OFFICE OF "THE MINING JOURNAL,"

26, FLEET STREET, E.C.

—

1884.

PREFACE.

During my present visit to England, my father, Mr. Robert Symons of Truro, pressed me to write a condensed account of the Geology of Cornwall, to accompany his forthcoming Gazetteer of Cornwall. This, owing to the limited leisure at my command, has been hastily written, and some of the matter may express my views imperfectly. Mindful of the diversity of opinion amongst the highest authorities, respecting the formation of veins, the author puts forward his ideas tentatively with a vivid consciousness of the difficulties which beset the elucidation of their genesis.

This reprint from the Gazetteer of Cornwall, has been made at the solicitation of many of the authors friends, who considered it desirable that the geological sketch should be published in a more readable form than that afforded by a large book of reference. Though a revision would have been most desirable, it was rendered impracticable because of the departure of the author from England.

In the method of treating so extensive a subject as the geology of a county endowed with such vast mineral riches as Cornwall, the author's object has been to avoid as much as possible, confusion of arrangement, and the use of technical language, and each subject will be found

as much isolated as the plan of the work would admit.
The author has desired especially to show that the mineral wealth of the county is very remote from exhaustion,
that, on the contrary, capital intelligently and honestly
applied to the development of ground, skilfully chosen,
would result in an increase of prosperity for Cornwall
and a fair return to investors.

To illustrate the effect that the secular denudation of
the diverse geological formations has impressed on the
physical character of the scenery, eight steel plate engrav-
ings have been kindly furnished by Messrs. Besley & Sons,
of Exeter. The geological map has been reduced from
the old Ordnance Survey, and the vein systems obtained
from the plans published at various times by Messrs.
Robert Symons & Son; whilst, for the colouring of the
Silurian strata, the author is indebted to the courtesy of
Mr. J. H. Collins, F.G.S.

In the preparation of this Geological Sketch, the
following works have been consulted : De la Běche's
Geology of Cornwall; Trans : of the Geological Society
of Cornwall; Moissonet's Rich Parts of Lodes; and
Collins' Minerals of Cornwall. Nearly the whole of the
statistical matter has been derived from the publications
of the Mining Record office.

Scottish Club, W.

CONTENTS.

ILLUSTRATIONS.

ERRATA.

Page 12, *for* archœan *read* archæan.

    ,,   20, *for* plentomaria *read* pleutomaria.

    ,,   23, *for* and the few beds termed greenstone *read* and some of the beds.

    ,,   67, *for* and the reverse of the dip *read* and the reversal of the dip.

    ,,   90, *for* sholes *read* shales.

    ,,   99, *for* tungstate of iron converted *read* was converted.

    ,, 104, *for* multation *read* mutation.

    ,, 174, *for* from the Saxon lodes *read* from the lodes in Saxony.

    ,, 178, *for* slime separations *read* slime separators.

GEOLOGY OF CORNWALL.

CHAPTER I.

HISTORICAL.

Cornwall is probably the most ancient mining country in Europe, if not in the world, and without doubt the Danmonians who worked the detrital deposits of tin on the granite bosses of Dartmoor and Bodmin Moor, during pre-Phœnician times, with which to fashion their domestic utensils, were the earliest tin miners. During pre-historic British ages the Phœnicians—the most enterprising merchants the world ever produced—crossing the Oceanus Britannicus were rewarded by the discovery of the stanniferous peninsula, which they called Belerion. A very important trade in tin, and perhaps lead, was developed, and for a long period Cornwall supplied the east with this valuable metal in exchange for brass, earthenware, and other articles, and consequently attained a civilization infinitely superior to the other provinces of Britain.

TOPOGRAPHICAL.

The peninsula of Cornwall is roughly shaped like a sea boot. It extends from the county of Devon, 80 miles in a south-westerly direction, to the Land's End, 30 miles

A

to the south-west of which are the Scilly Isles ; the width
varies from about 6 miles at Hayle, to 45 at the Tamar,
a large river which separates it from Devonshire. The
area has been computed at 1400 square miles, equal to
896,000 statute acres. A glance at the map will show
that the county is nearly surrounded by the waters of the
Atlantic Ocean, whose waves, born far out in its bosom,
lashed into fury by the violence of the prevailing westerly
gales, break with irresistible power on its shores. The
existing configuration of the Cornish coasts is indeed the
result of ages of marine erosion, evinced by the deep bays
and estuaries which have been hollowed out of softer
strata, whilst the compact rocks protrude as high craggy
headlands, whose obstinate resistance to the tidal surf is
clearly indicated by islets or chains of rugged reefs, the
most dangerous of which are marked by bells or are sur-
mounted by tall lighthouses, in which the keepers are
isolated for months by the unceasing roll of the Atlantic
billows.

Although the capes on the north coast are not so prominent as those in the English Channel, the coast line,
owing to the exposure of the edges of alternating dark
and white schist of unequal hardness to the nor-westers,
are much more indented, and numerous bold promontories,
bare and serrated, lend a character of grandeur and sub-
limity to the whole line of coast, unequalled by any
other portion of the English shore. The cliffs are everywhere so magnificent in diversity of contour, changefulness of the hues of its red brown, yellow, and blue rocks,
and in its variform steeps and precipices, that it becomes
difficult to localize the best points, but a pre-eminence
has often been claimed for the perpendicular columnar

granite of the Land's End, and the jagged slate rocks dominating the narrow porths in the neighbourhood of Tintagel.

Some of the islands, separated by oceanic action from the mainland, possess considerable historic interest. St. Michael's Mount must especially be noticed, on account of its admitted association with the ancient tin trade between Cornwall and Phœnicia. It is a granite cone rising about 240 feet above the sea, situate in front of the town of Marazion, from which it is separated by a shingle beach dry at half-tide.

Cornwall is surrounded by a shallow sea seldom reaching 40 fathoms, and this depth is only found near the principal headlands, where the tides rushing around them at a speed of four or five miles per hour, has swept out a clear channel for itself. The bottom slopes westward at so slight an angle to the edge of "soundings" that it would resemble, if raised to the surface, a boundless plain. The tidal wave arriving from the Atlantic flows up this slope, until striking the Land's End, one part expands itself in the broad waters of the English Channel, whilst the other, running with tumultuous rapidity into the Bristol Channel, becomes pent up, and by its reflex action causes an abnormal rise in the tide along the northern shores, the waters along which have thus a mean height somewhat in excess of that found to obtain on the southern coast.

The deep bays from the Land's End to the Rame head, and the less embayed shores on the north, are fringed by exquisite beaches of golden sands commonly made up of small comminuted shells, whose low specific gravity suffers the frequent gales which blow from the north-west, to

carry the sand up the cliffs, where its fall over the fields
spreads desolation. These sand dunes advance into the
interior until stopped by a small stream or a public road.
They are very extensive at Whitesand Bay, Heyl-mouth,
Perran, Newquay, and in the vicinity of Padstow. In
modern times the advance of the sand dunes has been effectually prevented by the growth of the reed called Arundo
arenaria. The numerous coves between St. Agnes and
Bude terminate in picturesque and romantic strands,
which are almost hidden beneath the high precipices that
render them more or less difficult of access. In the western coves quite a number of species of microscopic shells
have been described and figured; indeed, the beach consists
mostly of these, mingled with innumerable larger shells
and some quartz sand.

The force of the waves breaking against the shore,
causes a slow though continuous travel of the beach in
their direction, which is accelerated by the prevailing
current. Owing to the prevalent winds from the west,
the Cornish shingles move eastward, and this has led to
the formation of lakes at Helston and Swanpool, because
the volume of the fluvial waters became insufficient to
remove the accumulating sand and pebbles, so that the
final result is a wide bar which effectually separates the
sea from the fresh water surface behind it.

The heights of the county which form the water shed
are nearest the north coast, principally owing to the larger
river basins of the Tamar and Fowey. The narrowness
of the county precluding the existence of long river
courses, very many of the so called rivers are really mere
streamlets. The secular exfoliation of the granitic domes,
left divisional planes more or less open to atmospheric

action, which, softening particular bands, determined in numerous instances the direction of the rivulets that took their rise on the eruptive rock of the highlands. In the clay slates, the courses of the rivers appear to follow in many cases the junction of different strata, and this is more especially observable where formations of different ages are contiguous, Excepting the Camel, the rivers falling into the British Channel are insignificant, and no harbours of refuge for distressed ships are to be found ; but on the south littoral, the larger rivers expanding into estuaries, have given existence to commodious harbours and secure anchorage. Whilst all the southern creeks can be entered at any time, those debouching into the Bristol Channel, can only be made on the flood tide.

The peninsula of Cornwall, jutting far into the Atlantic, occupies a position extremely favorable to the fishing industry, and the waters of both channels are frequented during the season appropriate to the species, by immense shoals of mackerel, pilchards, and sardines, whilst hake, conger-eels, and pullock, swarm along the coasts. The headquarters of the pilchard fisheries is at Penzance, though St. Ives and Mevagissey are also important. Besides the fresh pilchards consumed in the county, from twenty to twenty-five thousand hogsheads of pressed fish are annually exported to the Mediterranean. The mackerel are sent principally to London, comparatively small quantities being sold in the county.

At Mevagissey, the young fry of the pilchard is canned and sold as Cornish Sardines. The oyster fisheries are nearly exhausted, though a certain quantity is still dredged at Falmouth and Helford. All the estuaries are frequented by salmon, and their peel are found far up the rivers, which are consequently much visited by anglers.

The heights of Cornwall do not extend in an unbroken ridge, but form a chain of elevated domes between which is low ground, though most of the north-eastern portion is raised considerably above the rest of the county. The physical appearance of a country depends very much indeed on the strata of which it is formed, and particularly when—as in the present case—eruptive and metamorphic rocks of diverse epochs are predominant. The central portion of the county is elevated from four hundred to a thousand feet, and is of a pre-eminently bleak and dreary character, especially during winter, when every object is shrouded in fog. The granite ranges, which are numerous and extensive, possess few dwellings to relieve the savage grandeur of its wild rocky carns and tors, nor are there any paths or tracks, by which a stranger could extricate himself from the deep and dangerous bogs which contour the hills and treacherously conceal their horrors beneath a smiling garb of verdant and luxuriant moss. The interior can only be trodden by the herdsman, who during the summer months leads his cattle to feed on the coarse grass and exuberant heath, whose purple hues clothe the hillside and invest the fantastic piles of granite rocks, which crown every summit with a splendour not all their own. The highlands where clay slate prevails are of a very much softer character, though sufficiently sterile. In place of tors and swamps, undulating and solitary plains, diversified in the north-east by hills of soft outline extend monotonously for many leagues. These hills, and frequently the downs, are disfigured by a layer of quartz or fieldspar, which imparts to them a dreary sterility much aggravated by the paucity of prominent objects to

relieve the uniformity. In summer these downs are be-
decked with tufts of variegated heaths, which in autumn
is succeeded by a gorgeous outburst of the furze blossom
that embues the gentle slopes with golden hues. Half a
century since, these downs or commons were far more extensive and desolate, but thousands of acres have been
brought into cultivation by the miners and small farmers
with incredible labour, worthy of more generous
leases.

Anciently these downs are said to have possessed a syl-
van character, and in many places remains of charcoal
pits have been ploughed up, and decayed tree stumps exposed in localities where no trees have been known to
flourish by the oldest farmers. In the mining districts
which usually occupy these commons, the barreness is inten-
sified by unsightly burrows of mine rubbish, which, intermingled with rough and rickety wooden structures carrying
tramways, flat rods, and hauling chains, with ragged sheds,
and the varied unclean appliances which the dressing of
the metallic ores necessitate, combine to form a scene, the
dismal wretchedness of which remains unequalled. Many
square miles of country are thus laid waste and rendered
worthless for agricultural purposes, though the industrious
miner, who is usually the proprietor of a house and plot
of ground, labours with infinite toil and commendable
perseverance to level the burrows in order to obtain pasturage for his cow.

The more or less circularly shaped granite bosses that
extend in a linear direction from Dartmoor to Land's
End, lose elevation towards the west, thus, the highest
tors of Dartmoor attain a height of 2000 feet, which
diminishes to 1368 at Brownwilley on Bodmin Moor,

1034 at Hensbarrow tumulus, 850 at Crowan Beacon,
and 800 feet at Carminnis, until at the Scilly Isles (Hes-
perides of Herodotus) the highest granitic peaks are but
220 feet high at St. Martin's obelisk, the rock finally van-
ishing beneath the ocean in an extensive reef. The trunks
of forest trees, often found in the detrital accumulations
resting in the depression of the eruptive bosses, seem to
imply that at a remote period the heights were clothed
—partially at least—in forest growth, however this may
be, the present violence of the winter storms scarcely permits the growth of trees on the highlands of the western
part of the county, where they are exposed to the full
force of the Atlantic gales.

The few stunted trees and straggling bushes striving
to vegetate, indicate unerringly the direction of the prevailing winds by the growth of their branches eastward.
This deficiency of forest land, has developed a large tim-
ber trade with the Scandinavians to supply the pine wood
required to keep open the shafts and levels of the numerous mines.

The **Scilly Islands** are composed wholly of ordinary
granite, and consist of six large islands and a multitude
of islets and rocky masses to the number of a hundred
and forty-five. Numerous rock basins, many of large
size and by some believed to be Druidical, which crowd
the surfaces of large blocks, are hollowed out by the action
of rain water. The scenery amongst the islands is romantic, though the want of trees much impairs its beauty ;
they are adorned by bushy heaths which the semi-tropical
character of the climate has caused to cluster luxuriantly
amongst the surface rocks. Palms may be seen flourishing at Tresco, and the geranium flower ornaments with a

flood of crimson the fronts of the cottages and the garden
fences. Fish multiply amid the lovely rocky isles in se-
curity, and the western reef literally swarm with gigantic
seals, while during the mating season myriads of sea-fowl
flock around the rocks above, or cover acres of the surface
of the adjacent sea. Coral grows abundantly not only
here, but in the vicinity of Falmouth.

If the scenery amongst the downs and moors fails to
excite the softer emotions of the heart, this cannot be
said of the beauteous and fertile valleys, which, taking
their rise amongst the tors, course southward through
charming dells, woods, and flowering prairies, to the English Channel. No description would do justice to the
varied and delightful landscapes which abound on the
southern slopes.

In the time of the Romans two **Roads** were made, one
on the north through Stratton, Bodmin and Redruth, to
St. Ives; and the celebrated Watling street extending from
Saltash through Liskeard, Truro, and Helston to Penzance.
These roads are still the chief thoroughfares, and whilst
the latter passes through the most picturesque scenes of
the south, the former traverses the grander regions of the
Cornish heights.

The **Climate** of Cornwall is varied and uncertain; on
the north coast the atmosphere during summer is bracing
without being cold, whilst on the south it is humid and
relaxing, and westward, at Mount's Bay, the summer is
very hot, and at Scilly becomes even sub-tropical. The
winter is nowhere severe, snow seldom falling, but the
continuous rains and mists make the climate during this
part of the year very trying to all but those of robust con-
stitutions. The mildness of the climate compared with

that of eastern and central England is due to its more pronounced insular character, and to the heated waters of the Gulf Stream. This maritime situation ensures an equable temperature, and the meteorological register shows but a small range.

The population is variable on account of the irregular state of the metal markets, because during times of depression the miners wander abroad in search of employment. Just at present the county is suffering severely from the low prices which prevail for metallic products. In 1871 the total population amounted to 362,098 and this compared with the last census shows a decrease of inhabitants of 32,614.

The Land's End.

SEQUENCE OF GEOLOGICAL FORMATIONS.

Rarely does such a small tract of country as Cornwall, possess such advantages and incentives to the study of vein systems associated with primary schists that have been mineralized by the upheaval and intrusion of eruptive rock. For the scientific classification of the strata, and the minerals they enclose, the Geological Society of Penzance was originated in 1818, since which time many eminent geognosts have, with unselfish zeal, laboured to adjust the chronological position of the Palæozoic beds of the west, and to elucidate the genesis of the innumerable fissures, by which they are disrupted. Glancing over the map issued by the Geological Survey, the colouring represents a chain of granite islands rising above a sea of slates. This simplicity of geologic structure is, however, only apparent, the rocks when studied minutely presenting complications which the scanty opportunities for investigation afforded by a cultivated country, render difficult of solution. The similarity of the Cornish killas, aggravated by metamorphism, cleavage, and contortions, together with the consequent broken condition of the often scarce fossils, adds to the difficulty of clearly determining the relative age, or defining with anything like accuracy, the precise superficial extension of these ancient formations ; indeed, but for the length of the coast line, with its lofty precipices and deep inlets, the sequence of the clay slates west of Liskeard would be still more ambiguous than is at present the case. Allowing for the irregularities produced by upheaval and denudation, the clay slates become gradually older from the carboniferous beds of Launceston and Bude, to strata below Silurian in the west. The formations in this series are shown to be distinct by their unconformability, and by the dissimilar strike of the bedding, but whatever position they may hold in order of time, they alike exhibit plenteous evidence of contortion and denudation, each having furnished the materials for the building up of its successor; moreover, each has been disturbed by the intrusion of igneous rocks before the upheaval which produced the granitic domes.

According to Mr. J. H. Collins and others, who have devoted themselves untiringly to remove the obscurity which has so long enveloped the age of these primary schists, the following table represents very nearly the stratagraphical order of the Palæozoic strata.

Formations.	Rocks.	Strike of Strata.	Dip of Strata.	Thickness of Strata.
				Feet.
Archœan ?	Mica slate, &c., of Lizard ..			
Cambrian ? ..	Ponsanooth Beds	N.W.	N.E.	12,000
Lower Silurian ..	Veryan Beds ..	S.W.	S.E.	23,000
Upper Silurian ?	Fowey Beds ..	S.S.E.	E.N.E.	10,000
Lower Devonian	Liskeard Rocks	E.	S.	
Lower Devonian	Plymouth Rocks	E.	S.	
Upper Devonian	Petherwin Beds	E.S.E.	N.N.E.	
Carboniferous ..	Launceston Beds	E.	S.	

ARCHŒAN. The Lizard Head is composed of micaceous and talcose slates, &c., similar in character to those on which the Eddystone lighthouse is built, and to the large extent of mica schist forming the southern ex-

tremity of Devonshire between Bolt Tail and the Start
Point. These rocks form portions of a metamorphic series
bearing little relation to the Cornish strata, which were
subsequently deposited. Around the serpentine of the
Lizard promontory are some highly crystalline metamorphic rocks, which are considered by some to be of pre-
Cambrian or Archœan age. They consist of gneissose
rocks, hornblende schist and rock, gneiss, gabbro, and
diabase, in nearly all of which the hornblende has a dis-
tinctiveness and brilliancy rarely observable elsewhere.
The constituents of these varieties are hornblende and
felspar, generally in about equal quantities. They appear
to dip under the mass of serpentine, though this is not
satisfactorily determined, because on the north in certain
localities they dip from it. The junction of the clay slate
with the hornblende schist is signalized by the presence
of a conglomerate, which is sometimes besprinkled with
native copper and some mundic.

CAMBRIAN. Excluding the metamorphic zones
of slate mantling around the eruptive bosses, and those
portions cut up by numerous granitic dykes and lodes,
the clay slates lying at a distance from the granite are
not much altered from their original condition after consolidation.

The oldest slates sweep around the Carnmenellis granite
from Godolphin through Crowan, Gwinear, Camborne,
and Redruth, to Perranwharf and old Kea, extending
also northwards to the Illogan and St. Agnes cliffs. On
consulting the ordnance map, this formation is found to
embrace the richest and oldest tin mining districts in
Cornwall. There seems little doubt that these beds are
pre-Silurian, on account of their ancient appearance and

rough foliated and siliceous character, and Mr. Collins
has classed them provisionally as Cambrian. They enclose
neither sandstone nor limestone, and no fossils have ever
been found in them. Throughout the whole formation
the basset edges of the beds have a north-westerly strike,
with a dip at Ponsanooth close to the granite of 45° to
the north-east, though on following the strata eastwards
this angle becomes reduced to twenty degrees. The total
thickness of the formation may reach 12,000 feet. Similar rocks rise from beneath the Lower Silurian near
Cargoll and come out at Penhale Point; hard silicious
rocks are also visible on the coast south of Pentewan, the
strata in both cases having a direction similar to that of
the principal mass.

The rocks of the **LOWER SILURIAN** occupy a
much larger area than those of the Cambrian, and the
included fossils leave no doubt as to their geologic age.
This formation is much intermingled with that of the
Devonian, the upheaval and denudation of the latter
having exposed in an irregular manner the Silurian strata.
They surround in a broad belt the Hensbarrow granite,
extending northwards through St. Wenn and St. Breock
to Wadebridge and enclose the mines of St. Austell and
St. Enoder. There is also a patch covering parts of the
parishes of Perran, Cubert, and Newlyn. The whole of
the Gwennap and Chacewater mines are encased in Silurian strata. They also stretch from Mevagissey and Ladock along the south coast to Helford, occupying both
sides of Carrick Road. All the clay slates between the
serpentine of the Lizard and the granite of Carnmenellis
and Tregoning, and the whole of the schists west of Camborne to Penzance and St. Just, are considered to be of

Lower Silurian age. The mining fields of Wheal Vor, Marazion, Hayle, St. Ives, and St. Just, find themselves in this formation. A very large proportion of the slates consists of rather hard thin bedded rocks usually grey or blue in colour, which sometimes assume a roofing slate texture, or deteriorate into tender arenaceous shales. Amongst these are numerous beds of highly siliceous rock that bear a considerable resemblance to intrusive greenstone, some thick bands of quartzite, and many siliceous conglomerates.

The strike of the strata is very constantly to the south-west, with a dip towards the south-east. Mr. Collins assigns a thickness of 23,000 feet to the rocks of the Lower Silurian.

The UPPER SILURIAN rocks usually spoken of as the Fowey beds, occupy a limited area between the Cornwall railway and the coast, along which they prevail from St. Austell Bay to the waters of the Hamoaze. Fossils, especially icthyolites, are plentifully found along the littoral in the arenaceous beds, and fix with some certainty their stratigraphical position. The prevalent colour of the slates is a dull grey or brown, which shades off into yellow and lake.

The beds are but little disturbed, and there is a characteristic absence of greenstone, with the presence of some beds of limestone. Although Restormel iron mine and the Herodsfoot lead mine have been worked in these rocks, they cannot be esteemed as highly metalliferous. The strata have a regular strike to the south-south-east, with a east-north-east dip of 26° for several miles, and this has enabled Mr. Collins to assign to the Fowey beds a depth of deposition equal to 10,000 feet.

The Devonian rocks, which rest unconformably on the Silurian, spread over all the eastern part of the county as far north as Launceston, where they descend beneath the carboniferous formation.

The **LOWER DEVONIAN** is well developed in the parishes of St. Austell, Ladock, and St. Allen, extending along the shore from the Gannel to the farther side of the Padstow estuary. The Ladock beds consist of alternations of hard dark grey and bluish schists, with soft reddish or yellow slates, and with some conglomerates and sand stones that course with remarkable persistency and regularity from Mevagissey to Perran. Notwithstanding that these beds are much disturbed by intrusive rocks, and are folded into anticlinals, they have a regular east and west strike and a general southern inclination. In the Ladock beds, throughout the whole thickness of 1500 feet, neither limestone nor fossils have yet been observed.

To the south of Bodmin Moor in the neighbourhood of Liskeard, are Devonian rocks that are believed to be the lowest of the series. They extend eastwards across the higher tidal waters of the Tamar, and the largest part of the Caradon and Tavistock mining districts are found in them. The Plymouth red sandstone and slates overlie the Liskeard beds. Both these sub-formations have a regular east and west strike, and, though there is some crumpling of the strata there is a prevailing dip to the south.

The **UPPER DEVONIAN** rocks, termed also the Petherwin beds, which border the Carboniferous formation, reach to the foot of the granite ranges of Bodmin Moor and Hingston Down. They enclose throughout large bands of greenstone or altered siliceous slate, intercalated in hard fine argillaceous beds, which, curving around with

Gneiss and Hornblende Rocks. — MULLION, near the Lizard.

Pub. by H. Besley Directors 20. 20 South St. 20 South St. Exeter

the granite, become so highly crystalline to the north and west of Camelford, that excellent roofing slate is obtained from them. The strike of the beds seem to be influenced by the contour of the granite, from which they incline, and distinctly dip under the shales of the next formation. The Upper Devonian rocks of Cornwall are but little metalliferous, only a few superficial deposits of manganese having been worked.

Only an unimportant area of the extreme north-east of the county is occupied by the grits and shales of the CARBONIFEROUS system, which repose with considerable lack of conformity on the slates of the last formation. They have been much disturbed by igneous action, though no ores of metals other than a few manganiferous deposits of little magnitude have been discovered. The general strike of the beds is east and west, and they frequently have a southern inclination.

Although there are no rocks of the CRETACEOUS epoch nearer than the east of Devonshire, a considerable quantity of chalk flints and fragments of the Greensand lie scattered over West Cornwall. They are especially noticeable on the eruptive rocks of the Lizard and Land's End, and even on the granite of the Scilly Isles. No satisfactory explanation of their presence has yet been suggested, but it has been imagined that the formations containing these flints may have extended farther west than they now do.

The whole of the county is very destitute of limestone, none existing west of Truro, and the few limestones occuring to the east being scarcely fit to burn; consequently the lime used in Cornwall is quarried in the extensive and excellent calcareous strata on which

B

Plymouth is built. Such beds as exist, usually crop out
from the argillaceous slates in thin and irregular beds of
very inferior quality, having little extension either
superficially or in depth. They enclose numerous
fragments of fossils, which on account of the altered
character of the lime rocks, are only visible in thin
sections when examined through the microscope. Calcareous beds occur in patches along the south coast from
Helford to the Hamoaze in Silurian strata, and on the
north coast from the Gannel to Padstow estuary, and
also in the neighbourhood of Petherwin, where it is
intercalated with Devonian beds.

The conditions under which the Palæozoic slates of
Cornwall were accumulated, were little favourable to the
existence of those organisms on which the deposition of
calcareous matters is dependable. All the Cornish strata
are more or less fossilliferous as far west as Truro, after
which they become involved with the granite dykes, and
so disrupted by fissures, that metamorphism has probably
destroyed any fossils which may have previously existed.
The Cambrian has been found very barren of fossils, for
though much search has been made in them, only rare
fragments of tabulate corals have been discovered. The
Lower Silurian beds are more productive, and many of
the fossils, especially **Orthidæ,** are in a better state of
preservation. In the neighbourhood of Bodmin are
Trilobrites and crinoidal remains. The quartzites of
Gorran and Veryan are sometimes very prolific, particularly at the Nare Head, where **Orthidæ** abound,
and **Terebratulæ** and **Trilobites** are found. West of
Truro fucoidal impressions have been noticed.

Though the superficial extent of the Upper Silurian

beds compared to that of the older subformations is but small, yet it is distinguished by a large number of interesting fossils, representative of many of the Palæozoic marine groups.

The petrified organisms are found most numerously along the littoral from Whitsand Bay by Polruan, Fowey, and Par, to Pentewan. At the Black Head near the last named place, the discovery of **Graptolites** have definitively fixed the Silurian age of those rocks. From the Rame Head to St. Austell Bay, corals have been abundantly discovered, and in many places as at Crennis, at Fowey, and in the limestone of the Black Head, are numerous remains of corals and encrinal stems, together with traces of other organisms. **Trilobites** are found also at various points, as well as **Brachiopodæ.**

The Fish beds that prevail from Polruan to Whitesand Bay are indicative of Upper Silurian age, though eastward they are said to approximate to the Lower Devonian group ; the species however being peculiar to the former. The fragmentary state of these fish bones render somewhat doubtful the genus **Onchus,** which has been based on the variety of defensive spines found.

The beds of the Lower Devonian enclose many fossils, particularly in the vicinity of Liskeard, and between Newquay and Padstow, where corallines, encrinites, and similar genus abound in the beds containing calciferous matter, that prevails between the Hensbarrow granite and the Bristol Channel.

Trilobites, numerous in Roseland valley east of Liskeard, are rarer on the north coast ; they are much broken and often even comminuted.

The large Sub-Kingdom **Mollusca** has one or more

genus in every class represented by fossils. Thus, amongst the Brachiopoda, **Terebratulæ,** and **Spirifers** are not uncommon on the north, and these, together with **Producta, Cirrus,** &c. are often gathered from the Plymouth beds, the fossils of which are believed to be mostly referable to the Mountain Limestone.

In the Lamellibranchiata, **Bellerophon** is met with in the Liskeard beds and elsewhere; and among the Gasteropoda, **Buccinum** and **Plentomaria** are not uncommon, whilst **Orthoceratites** and **Goniatites** stand for the Cephalpoda. It is to be regretted that only a small proportion of these fossils are well preserved.

The beds of the Upper Devonian are so much metamorphosed by the proximity of the Bodmin granite, and by the numerous and broad bands of greenstone, that the fossils found are far from numerous, and usually in a state very ungrateful to a collector. They prevail, however, throughout all the beds which curve around the highlands from Tintagel through Petherwin to the Tamar north of Hingston Downs. The fossils found, omitting the **Trilobites,** are much the same as those in the Lower Devonian, and therefore need no recapitulation, but they represent more nearly the character of the Mountain Limestone, and vegetal impressions are common.

In the roofing district in the vicinity of Tintagel, the fossils have a defined outline, and are encrusted with gleaming green plates. Fragments of **Spirifers** are frequent in the quarries of roofing slate, and many of the fossils are singularly distorted through the elongation caused by the cleavage pressure that gave to the fine argillaceous slates their crystalline character.

GREENSTONE. The Devonian schists in the

north-eastern part of the county, and the slates between
Hayle and Mount's Bay are associated with numerous
groups of hard green rocks, which have been indiscrimi-
nately termed greenstones. They have not, however, the
same origin, but owe their existence both to intrusion
and metamorphism. By much the larger proportion of
these igneous looking rocks are found interbedded with
the slates, from whence, owing to their indurated
character, they often rise in low rounded knolls, or run
into stony eminences. The contemporaneous formation
of these beds appear clearly indicated by the coincidence
of their direction with the strike of the strata, whether
that be S.W. as in the Lower Silurian, east and west as in
the lower Devonian, or S.E. in following the superior beds
of the Old Red Sandstone. There is however an important distinction to be made in considering these rocks,
for while most can without hesitation be referred to
intrusion, many of the hard siliceous beds found in the
metalliferous districts, many owe their durability and
appearance to metamorphism. It is not always easy to
distinguish these two classes when they are proximate to
the granite, around which, on account possibly of the
upheaval of the whole mass of strata, and the subsequent
exposure of their basset edges by denudation, they
sometimes seem to contour. Though the bulk of the
TRAPPEAN ROCKS, especially those skirting
the granite, remain hard, they are by no means always
so, but often assume an amygdaloidal, vesicular,
pumiceous, cindery, or even earthy form, where decomposition has loosened the structure. All these beds of
greenstone, whether felspathic or siliceous, must have
been existant prior to the granite upheaval.

Greenstone is rarely visible amongst the Cambrian beds, though some intensely hard siliceous rocks are found above the edge of the Gwennap and Penryn granite. Similar bands, though more quartzose and thinner, are seen north of Redruth. In the vicinity of Penzance there is a large development of altered slates and contemporaneous trap rock, which is very interesting because of their intimate alternation with argillaceous slates, and the conformity of the strike of the bedding with the contour of the deep indentation of the granite. Hard greenstone rock fringe the granite from St. Just to St. Ives, imparting to the western cliffs the ruggedness and romantic grandeur which attracts so many tourists. In some of these rocks, as at Botallack, Lariggan, and other localities, actynolite and axinite in beautiful crystals replace the hornblende, and also occur in veins. Numerous beautiful greenstones are interbedded with argillaceous slates east of St. Michael's Mount, and can be well studied along the Perranuthnoe cliffs; also in the Silurian slates near Helston, and at Roskreage Beacon on the borders of the hornblende schist of the Lizard.

The bands of Irestone, which appearing in St. Erth, cross the parishes of Gwinear and Camborne to the Roskear mines, have long been known; they are harder than any other beds in Cornwall. They are composed of minute granular hornblende, compact felspar, and quartz. Few greenstones have been observed in the central part of the county between the granitic domes of Carnmenellis and Hensbarrow, and from the latter westward to the coast, and northward to Padstow inlet. The Upper Silurian strata from Fowey eastward are equally undisturbed by intrusive rocks, and the few beds termed greenstone in St. Stephens are merely altered slates. At St. Breock Downs is a remarkable group of quartzose bands, narrow, but well defined, whose strike and dip seem parallel to those of the accompanying slates for several miles. On the north of the tidal portion of the Camel are numerous contemporaneous greenstones, often vesicular, and similar rocks of Lower Devonian age east of Liskeard, have an east and west direction. The trappean rocks near the latter town exfoliate and leave hard nuclei like the greenstones of Saltash. The green-stone rocks of the Upper Devonian have a more important development than those of the more ancient formations, and appear to have been far more displaced by the granitic upheaval. The trappean rocks intercalated with the argillaceous slates, which have a strike a few degrees west of north in the parish of St. Teath, appear to take a bend around to the N.E. by Delabole and Tintagel, and continuing the curve arrive at the opposite side of the Bodmin moor, where their course is parallel to that of the eruptive rock, which has a S.E. bearing. The bands are broad and continuous for a long distance, and consist of every variety of greenstone, from the dark foliated hornblende rock found near the granite, to pumiceous and earthy beds stained with iron.

Besides the greenstone beds and contemporaneous trappean rocks, there is another class whose formative epoch is not so evident, as it cuts through the slates in irregular veins resembling the intrusions of granitic matter at the junction of that rock with the clay slates. Most of these are doubtless of pre-granitic age, though some may been injected during the cumulation of the

Lizard serpentine. The small and numerous trap veins
that find themselves near Newquay, Saltash, St. Cleer,
Helford, and at other places, which are filled sometimes
with vesicular matter, but generally with green or
greenish brown compact and homogenous rock, capable
of receiving a high polish, scarcely merit the title of
TRAP DYKES. The comparatively low temperature
which existed in the veins at the time of their intrusion
would be little likely to cause any considerable metamorphic effects on the enclosing rocks, and therefore,
though the slates in their immediate neighbourhood are
often displaced and even contorted, yet only the walls
of the larger fissures, with some included fragments,
have been here and there altered by heat, and by subsequent decomposition.

The siliceous greenstones and trappean beds and dykes,
though occurring through the whole Palæozoic series of
Cornwall, have not perhaps been observed with the
intentness that their intimate association with the best
metalliferous regions requires.

SERPENTINE. Though many parts of the
county are noted for the production of rare minerals, no
district possesses such geological interest as the Lizard,
where a group of crystalline rocks, consisting of
serpentine, diallage, and allied species exhibit a com-
plexity of detail, and an association of beautiful earthy
minerals rarely observed. The whole mass occupies the
southern portion of the promontory, and is nearly
surrounded by a broad interrupted border of hornblendic
schists, that appears to form a sort of basin enclosing the
serpentine plateau, whose limits are defined with
sufficient exactness by the beautiful Cornish heath,

Pub. by H. Besley, Drawing Room Table Books, S.W.

Serpentine Rocks. —— KYNANCE COVE.

called **Erica vagans,** which enlivening with its hues
the sombre solitudes of the spacious downs, is found
nowhere else except on magnesian rocks of analagous character, near Menheniot railway station.   The eruptive
period of the Lizard rocks is not yet conclusively
determined ; indeed, the present extent is due to more
than one bursting forth of igneous matter.   De la Bêche
imagines that they may have appeared during the
deposition of some of the Cornish Killas, though no
traces of serpentinous rocks have been discovered in the
conglomerates associated with those slates.   Its pre-
granitic age admits of little doubt, as it rests apparently
on granite, veins of which are found penetrating it at
many places along the encircling shores.   The serpentine
itself acts very much like granite, for though it seems to
have little disturbed the southern dip of the Helford
slates, there is, in some localities a kind of passage
between the hornblende schists and serpentine, with an
outward dip, and it has even broken through the slates
at Pencarrack, and cut through the hornblende schists at
Porthalla.   These junction rocks are accompanied by a
development of a limy substance, and much red colour in
the serpentine.   The Lizard rocks have a composition of
remarkable variety, accompanied by a rich diversity of
colours, that on polishing discloses a wealth of brilliant
shades of red, brown, green, and grey, whose silky and
pearly lustres are worthy of all the commendations
which its manufacture into artistic ornaments has
elicited.

DIALLAGE ROCK.   The diallage rock which
covers four or five square miles between St. Keverne and
Coverack, is due to an eruption posterior to the consoli-

dation of the serpentine, because many veins of diallage containing fine crystals of saussurite traverse it at Coverack Cove.   At Porthoustack and at Ruten Point, it has cut through greenstone and greenstone porphyry.   The rock is composed of felspar, hornblende, and diallage, so that while it is often true diallage rock, it sometimes resembles a coarse mixture of hornblende and felspar ; diabase is found and gabbro also.   After consolidation, fissures were formed in both serpentine and diallage rock ; a large course of diallage runs from Carakclews through Gwinter like an elvan, and similarly a dyke of trappean porphyry fissures the serpentine from Bochin southward to Penhale, whilst bands of trap rock prevail at Coverack Cove.   Veins of steatite are numerous, and were formerly quarried for the manu- facture of porcelain.   Unimportant veins, at the contact of diverse rocks, yield specimens of native copper, one of which, weighing a hundredweight, was sent to the exhibition of 1851.   Nowithstanding the extent of the magnesian rocks, chrome ore has only been noticed in minute quantities.

Serpentine in areas of small extent has been found in other parts of the county, as at Nare Head in Veryan Bay, where there is a group of rocks precisely like those of the Lizard.   There is also a small outbreak through the clay slates at Duporth, and a fine dyke at Clicker Tor exposing variegated shattered rocks, with steatite or asbestos in the joints.

THE GRANITE ROCKS.

It has been shown that the clays, sandstones, con- glomerates, and limestones which surround the eruptive

highlands of Cornwall, though all included in the Palæozoic or Primary group of formations, have yet been deposited at periods so immensely removed one from the
other, that there was time sufficient for a specific and
even generic mutation of marine life, and for a change
in the deep seated axes of upheaval, by which some of
the clay slates were folded into ridges, from which the
strata inclined both ways.   Thus the portion of Britain,
now called Cornwall, was sea and land alternately before
the upheaval of granite, which was to transform a mass
of barren schist into rocks teeming with metalliferous
wealth.   Some physicists are of opinion that the granite
range, stretching from Scilly to Dartmoor, is the result of
upward thrusts given at distant epochs by forces acting
in diverse directions; but geologists who have devoted
their time and abilities to the study of the Cornish strata,
consider them as formed during the same geological
period, and place the epoch after the deposition of the
Upper Devonian.   From the constitution of the granite,
it is inferred that it was cooled under a pressure, whose
equivalent has been calculated at five miles of depth.

The general direction of this granite ridge is east-north-
east, and though no doubt the original thickness of the
slates brought up on the granite was very great, during
and since emergence from the ocean, denudation has
exposed a chain of granite domes of various extent, and
reduced the thickness of the surrounding slates so considerably, that over most of the county the backs of the
elvans and lodes have been brought within workable distance of the miner.   So much appears to have been swept
off some of the granite ranges, that the nucleus is approached, and the manner in which they were built up is

thus rendered to some extent obscure, but close observation of the numerous masses and broad bands of rock differently aggregated, supplies strong evidence that the earlier consolidations were subject to numerous disruptions, which not only elevated the mass, but extended it laterally. The granite produced by secondary eruptions, may perhaps be recognized by the corrugated surface of the granite from Carnbrea underground to East Pool, by similar ridges north of Hensbarrow, by the heterogenous aggregation of the composing mineral, and variety of texture assumed by broad bands of rock. Most of the subordinate intrusions have a direction more or less co-incident with the general trend of the range, but frequently they are small, curved, and irregular. The smaller tracts are fine grained, though often coarse when of great width, but both are distinguished by a finer and harder texture along the margin. Veins of greisen are not rare, especially in decomposed granite, though many of them may be considered lodes. Hensbarrow granite, which has possibly suffered less denudation, has a texture and composition of remarkable inconstancy, and is traversed by veins of finer or coarser texture, which in rare instances are nearly " giant granite."

The granite protrudes above the slates in masses of rounded form, and the exfoliation produced by the reaction of external influences, has given rise to a massive lamellar structure which is very conformable to the surface and dips at gentle angles under the clayslate. This, combined with the fissures due to the divisional planes, has caused the granite to assume the columnar and tabular appearance so characteristic of this rock. It will have been perceived from the above remarks that the reiterated fractures sustained by the consolidating granite crust, could only have resulted in a very considerable diversity of the proximate constitution, as well as the texture of the erupted masses. The more closely the rocks are observed, the clearer becomes the evidence of the extreme incongruity of the granite. Generally the rock has a coarse granular appearance, which is most pronounced in the Land's End district, and is of the finest grain on Dartmoor. The composition, though subject to endless minor deviations, is usually a mixture of felspar, quartz, and mica, with some schörl ; the latter, though scarce in the central positions, often assumes much importance towards the confines of the granite. This is especially the case around the Hensbarrow boss, where even the rocks of the interior abound in schörlaceous veins. Large areas are rendered porphyritic by the occurrence of big crystals of felspar, sometimes white, but beautifully tinted in the vicinity of the metalliferous deposits by incipient decom position. As a rule it may be said, that the crystals forming the aggregate, are individually imperfect, or have their edges not sharply defined.

The granite is also drier, more compact, and of a more homogenous texture when far from its junction with other rocks. There are limited areas where, as at Godolphin Hill, the component crystals are embedded thickly in a felspathic base. In some districts, notably in the neighbourhood of St. Austell, the granite consists of felspar and quartz, in which the former, under the influence of schörlaceous veins, has become so soft as to admit of its being washed for china clay. The granite is so variable that it is quite impossible to give its chemical composition with anything approaching to accuracy.

If we take an ordinary porphyritic kind, in which the proportion of felspar, quartz and mica are respectively equal to about one-half, one-third, and one-sixth, there would be about 73 % of silica, 19 % alumina, and 8 % of potassa, besides fractional percentages of iron oxides, magnesia, lime, soda, manganese, and fluoric acid, and where schörl abounds, boracic acid.   Other common varieties do not contain so much silicic acid.

As granite owes half its bulk, and nearly all its mobility to FELSPAR, its character and resistance to decomposing influences is very dependent on that mineral. When the felspar weathers, the rock separates into its component parts, and the surface is strewed with growan. The felspar crystals bestow on the rock its ensemble, thus in the finer grained and non-metalliferous portions, the confused aggregation with felspar, gives the granite a dull grey colour, whilst in those places mostly approximate to the mining fields, where the felspar crystals are as a rule ill defined, and where incipient oxidation has altered their colour, variegated shades of brown, red, crimson, green, &c., impart to the rock characteristic hues that please the eye, and guide the miner in his search after subterrene ores.   The large felspar crystals are sometimes removed by solution and re-filled with binoxide of tin as at St. Agnes Beacon, or adorned by radiate schörl as at St. Ives. The felspar of Cornwall (orthoclase) is a silicate of alumina and potash, containing about two-thirds silica, one-fifth alumina, and one-eighth potassa, with a small admixture of peroxide of iron, lime, and soda.

The decomposition of the felspar in some of the granulitic portions of the granite, changing the latter to a soft crumbling mass, renders the potash and part of the silica soluble, and thus leaves the hydrous silicate of alumina, which is washed out and sold as china clay. Small tortuous veins of felspar frequently traverse the granite.

After felspar QUARTZ is the most important ingredient of granite. It is amorphous and seems to have been to some extent plastic, when the mica and felspar had assumed more or less their forms. It is generally pellucid, has a pearly white lustre, and often appears to have a smoky interior. In the midst of the granite ranges, the quartz is distributed with considerable evenness, but where disturbed, or when near the junction with other rocks, it is more variable, and contemporaneous veins of quartz, or quarts mingled with felspar, are frequently to be observed dividing the granite for quite a distance.

The MICA though usually small in quantity, has numerous, black, brown, and silvery white colours that give a pleasing brilliancy of appearance and diversity of character in the mineral districts, although more dull and monotone elsewhere. Short veins of mica are in some localities numerous, but are rarely of magnitude; half-an-inch in width or less being the common size. The following analysis of Cornish varieties will give a general idea of its composition.

	GREY MICA.	BROWN MICA.
Silica	50·82	40·06
Alumina	21·33	22·90
Ferric Protoxide	9·08	27·06
Oxide of Manganese		1·79
Fluoric Acid	4·81	2·71
Potassa	9·86	4·30
Lithia	4·05	2·00
	99·95	100·82

Schörl is widely diffused through the Cornish granites, but is generally in small proportion in the main masses, except in the bosses of Land's End and Hensbarrow. In the latter it plays a most important part, as also along the junction of the slates with the eruptive rock. The following is a proximate analysis of the black variety so common in the Cornish granites.

Silica	...	...	36·10
Alumina	...	...	34·29
Oxides of iron	...	...	16·69
Boracic acid	...	...	5·88
Manganese	...	...	·22
Magnesia	...	...	1·64
Lime	...	...	·54
Soda	...	...	1·74
Potash	...	...	·34
Fluoric acid	...	...	·74

$$98·18$$

Neither **HORNBLENDE** nor **TALC** are abundant in ordinary granite, though chlorite and talc replace to a small extent the mica in varieties; **PINITE** is however rather common in the Tregoning and Land's End district. At rare places, as in the Carn Marth rocks, fluor spar is found as a constituent.

Tin ore or cassiterite is found disseminated in patches of granite in St. Just, and on the eastern side of Tregoning Hill, but it can scarcely be said to be a constituent of the granite.

GRANITE VEINS. The eruptive matter which rushed into the earthquake chasms formed in the earliest consolidated granitic crust, was probably injected at the same epoch as the granite veins which rent asunder the adjacent sedimentary strata whenever the requisite pres-

sure became developed by the upheaval of the granite.
As granite veins are seen cutting the clay slates in small
tortuous veins, or penetrating between the beds wherever
a junction is exposed, either amongst the rugged cliffs or
in the mines, it is rational to infer that they occur along
the whole line of juxtaposition. The character of these
veins, which have no definite direction, is the same as
the granite from which they were derived. The veins
are usually of no great length, but since at Porthleven
Cove and other places, they may be seen less than an
inch thick, though of considerable length, the eruptive
matter must have possessed a considerable amount of
fluidity. From the numerous fragments torn from the
slate and isolated in the veins, the force of the eruption
must have been immense; in some places the molten
rock has forced itself through strata which have closed
behind it and left isolated masses of granite. Granite
veins gradually fine out, have no distinct walls like elvans
and indifferently intercalate with the strata or run
athwart them. Owing to some difference in the rate of
cooling, the smaller veins have more quartz and less mica
with a finer grain; and the larger, though crystalline and
even porphyritic in the middle, are compact along the
sides. All the veins and isolated masses of granite are
recognised, after due examination, to proceed from the
main mass, and like it are non-metalliferous. If one may
judge from the numerous granite veins visible about the
beach and cliffs of the Lizard, penetrating serpentine and
diallage rock; the inference that the magnesian rocks of
that promontory have also suffered from the same upheaval, and now overlie granite is difficult to resist. The
localities where the veins occur are much too numerous

to mention, but they can be best studied at Trewavas, near Helston, Polmear in Zennor, Porthleven in St. Just, and underground in the mines which are opened along the junction at the northern foot of Carnbrea Hill.

SCHORL ROCK. This rock is so intimately associated with the granitic rocks of Cornwall, that although many dyke like masses of schörl rock are found along the junction and running into the slate, and numerous schorlaceous bands and veins traverse the granite itself, it is quite impracticable to colour it on the map apart from the eruptive rock. The schörlaceous granite occurring in bands amid the ordinary granite, it may be that they were formed after the first consolidation of the erupted granite, though many alterations must have since taken place, owing to the facile manner in which schörl changes its locale; thus, it has been noticed that when schörl has penetrated into the cavities left vacant by the disappearance of the felspar, the enclosing rock in the immediate vicinity possesses little or none of that mineral. Many of the schörl rocks have a beautiful appearance; they are white when very quartzose; mottled, grey, and black in Luxulyan valley, where pinkish orthoclase and black schörl are embedded porphyritically in a grey crystalline quartz; whilst in another variety small prisms and groups of stellate schörl are thickly sprinkled through similar rock. The varying proportions of schörl to quartz are infinite, and while sometimes schörl is predominant, at others the quartz rock may contain scarcely any. Thus some species are even harder than the granite, and resist atmospheric influence far more successfully. This resistance is very pronounced in the western portion of Hensbarrow, where the Calliquoiter cairn, and the celebrated

Roche Rocks, break the monotony of the Moors, and invest them with sombre grandeur. Beside forming rugged ridges of schörl rock, this mineral is found abundantly amongst the altered rocks which so constantly accompany the junction of the clay slates with the granitic masses. The transition from granite to clay schists is remarkably interesting in many places, but perhaps nowhere is it so instructive as at Carclaze, where the lamellar varieties of schörl rock are exposed in a vast hollow that has been excavated across the junction, for the purpose of extracting the tin ore and the china clay accompanying the schörlaceous schists. As far as observation has gone, schörl is characteristically associated with tin ore amongst the altered slaty felspar beds along the margin of the granite in the famous mines of Tincroft, Cook's Kitchen, and Dolcoath. The unstable character of schörl, and its power of altering the appearance and character of both granite and slate, is shewn by its frequently occupying the joints in granite lodes, and by the occurrence of the tourmaline schists that gradually blends the granite into slates. But in addition to the prevalence of schorl in the positions above indicated, it plays also an important rôle throughout the entire mass of some granitic bosses, and this is most especially the case in the St. Austell hills, where schörlaceous veins are very numerous, and generally connected with the presence of both tin ore and kaolin. From the variety in composition and appearance of these veins, and the irregular manner of their occurrence, it is not an easy task to give an intelligible and concise description of them,* but the facility with which schörl

---

* Vide pamphlet published by Mr. J. H. Collins, F.G.S., on the Hensbarrow granite, where will be found a full and excellent account of these rocks.

rock passes into granite, gives colour to the suspicion
that a large portion may be simply altered granite, due to
the replacing of felspar by schörl; even brecciated masses
have been found re-cemented by it. Although in the
other eruptive bosses of Cornwall, schörl is not so prevalent, yet it may be said that schörlaceous veins—sometimes with felspar, sometimes with quartz—usually
associated with cassiterite, are found in the granite of
every mining district in Cornwall and Devon. Besides
those in the Roche district, large masses of schörl rock
are prominent at Wheal Trannack near Helston, Porth-
leden in St. Just, and at Tresavean in Gwennap.

CONCEALED GRANITE. Nearly all mining
districts are clearly recognised to bear a close and intimate
connection with the metamorphosed rocks, which the upheaval of the granite through the argillaceous schist has
produced. Though from a cursory examination of a
geological map of the county, the numerous lodes and
cross courses found far from the confines of the granitic
bosses, would appear to controvert this statement, a little
reflection will lead to the conviction that the mineralized
strata are almost without exception closely approximate
to the eruptive rock, if verticality be regarded. Taking
such districts as Marazion and Chiverton, both distant
some miles from the junction of the crystalline with the
schistose rock, we may feel assured from the numerous
elvans which traverse the slates, that the main mass of
granite is not distant. Though the varying size and dip
of these elvans modifies to a considerable extent their
porphyritic character, still from their appearance and mechanical composition, the relative remoteness of the nuclei
from whence they proceeded can be roughly estimated.

Taking for examples the thinning out of the Gwennap group in the neighbourhood of Truro, where the porphyritic structure is almost lost, and elvan acquires the texture of sandstone, and the Chiverton group, where, but for their occuring in slates they might easily be taken for sandstones, one need not hesitate to affirm, that where these granitic dykes are wanting, the argillaceous slates become barren when distant from the eruptive domes. In the mining districts rarely found on the granite, the proximity of the slates seems also to have been essential, and though it would be rash to attempt to judge the amount of denudation which the granite has suffered from the metalliferous character of the existing surface, still there is a blank barrenness of mineral in some bosses when compared to others, which taken together with the rich stanniferous gravels that once filled the moors, would lead to the conclusion, that the more or less want of metalliferous wealth may be in relation to the depth of crystalline rock removed by the agency of water.

It was formerly considered that the granite inclined under the clay slate at about an angle of 45°. This has since been found in several instances not to be the case, the most important exception, and the one best known, is that seen in the mines lying on the northern slope of Carnbrea Hill; here, the granite after reaching a depth or 115 fathoms commenced to rise again until the top of a ridge was attained, and although it renewed its northerly dip, it rose into another ridge at East Pool. There can be little doubt that this character is prevalent going north, for, on the coast of St. Agnes, and Perran, the granite again comes to the surface at Beacon Hill and Cligga Head. Notwithstanding that the observations of geolo-

gists have shown conclusively, that the Cambrian and
Silurian strata suffered from disturbance and contortion
before the granitic upheaval, there are many tracts in the
clay slate, where a quaquaversal dip of the beds, and a
domelike contour, suggests the existence of eruptive bosses
hidden beneath strata of moderate thickness. Thus the
elevated ridge of St. Breock Downs—with its dyke-like
courses of quartzose rock—consists of slates which con-
form with much accuracy to the swell of the hill. At
Camelford, where the visible granite ceases, the seaward
curve of the greenstones and their anticlinal dip, seem to
indicate that the granite would be found at no great
distance from the surface on the high ground between
Roughtor and Cadon Hill near Tintagel. The funda-
mental axis of the eruptive rock is plainly detectable at
the surface through most of the county, by the groups of
elvans that run out in the direction of the succeeding
boss. This can be clearly recognized between the mountains of Bodmin Moor and Dartmoor, where the elvan
dykes stretching across the fruitful valley of the Tamar
expand into the two small granitic eminences of Kit Hill
and Hingston Downs, midway between the mountains.
Though no main mass of granite comes to the surface in
the Lizard promontory, the appearance of small elvan
dykes, and the occurrence of numerous granite veins
around it, leave no doubt on the mind of the observer
that the granite supports the mass of serpentine.

KILLAS or CLAY SLATE. It has already been
observed that the slates of Cornwall belong to geological
formations deposited during most, if not the whole, of the
time assigned to Palæozoic times, and that the strata
composing them had been so disturbed, that in many

localities the beds were folded into anticlinals, and that
the whole series of slates, limestones, sandstones, and
conglomerates were consolidated, and had acquired to a
considerable extent their present structure and appearance
ages before the upheaval of the granite. The lentiles,
veins, and splashes of quartz found everywhere associated
with the schists, but most prevalent and contorted
amongst the older strata, show that silica in solution
prior to the granitic upheaval was not rare. Though these
veins are termed contemporaneous, because of their being
confined to a sequence of beds geologically limited, the
silicious matter has been deposited within them by percolation in the same manner as at present, and consequently
in the older and more disturbed beds, foliations and
masses of quartz are predominant. These veins are short,
irregular, and without walls; they have no persistent
strike, dip, or size, nor do they enclose any mineral.
The strata throughout the county vary in thickness
from that of a sheet of paper, to several feet, and though
usually clayey or fine grained, are often mixed with
arenaceous strata as in the Ladock beds, or with
conglomerate made up of moderately sized fragments, but
occasionally—as at the Nare Head—of larger pieces,
some of which reach the weight of several hundred
pounds. The beds lie parallel to each other, and though
sometimes uniform, are distinguishable by difference in
structure, colour, and texture. Exfoliation of the slates
along the joints and cleavage planes has much broken
them up near the surface, but below the influence of
atmospheric causes, they are sufficiently compact, and
the strata generally clearly defined. It must, however, be
admitted that the strike and dip of the clay schists are

often recognised with difficulty—especially in mineralized
strata—because they are so much obscured by the planes
of lamination, which are often the same as those of the
bedding, and sometimes much more prominent.   The
disturbance consequent on the upheaval of the eruptive
rock did not extend far from its borders ; on the contrary,
the distant beds—excluding the contortions due to
plutonic action before the granitic intrusion—in spite of
numerous changes of bearing and dip are but little deranged.

The activity of the plutonic forces, which awakened
after the consolidation of the Silurian and Devonian
formations, carried up the strata and left them arching
over the numerous domes which form the Cornish and
Devonian mountains.   It has been estimated that these
schists have had a minimum thickness of forty
thousand feet, and this thick mass must have been
denuded from the granite hills during its gradual rise and
emergence from the sea.   The clay slates like the granite
fail in height from east to west ; they mantle high around
the Devonshire tors, whilst at the Land's End they are
seen only in patches which disappear beneath the sea.
The crystalline schists now seen mantling immediately
around the bosses, are the denuded basset edges of the
strata uplifted by the granite, and must therefore form a
portion of the lowest beds of the formation to which they
belong.   These, metamorphosed by contact with the
igneous rocks, have assumed that compact crystalline, but
thick lamellar structure, which is so eminently characteristic of the metalliferous rocks, when in the immediate
vicinage of the granite.   The strata in the mineral
districts incline at easy angles ranging between 20° and
30° from the granite ; and this feature combined with

Granite Hills of Carnbrea in distance.

A TIN MINE

the similarity possessed by the altered slates, make it appear as if the same beds curved around with the granite, when really they often thin out, and are replaced by others almost identical in character.

The inclination of the slates is only high, when the line of juncture between the igneous and sedimentary rock happens to coincide with the strike of the killas, but though the mechanical effects produced, may seem to have been comparatively slight, the chemical alterations—to an important extent dependent on the original structure—are of the greatest magnitude, and have in an infinitude of ways transformed the contact rocks, and by replacement of one mineral for another, rendered it a matter of considerable difficulty to recognise their pristine character. It will not be deemed strange, that as the strata owe their changed condition to heat, their metamorphic character should be in direct relation with remoteness from the eruptive rock ; and accordingly it is found, that whether these be slaty greenstone, hornblendic, chloritic, micaceous, or compact felspar slates, at a distance from the junction comparatively small, a schistose clay slate diminishing in metalliferous value invariably succeeds. The contact rocks, both granite and slate, are usually hard and compact, though in some districts where schörl prevails, decomposition has softened them, and a sub-crystalline texture, with a massive structure distinguishes the slate beds. Their character is so diverse around the different granitic tracts, that only the principal varieties can be enumerated in this place, whilst any rock peculiar to a district, will be referred to farther on. The prevalence of schörl in the granite, and of mica in the slates is of general occurrence, and often

D

it is quite impossible to detect the true line of separation. When schörl is not present in quantity, the transition is often imperceptible, the granite becoming more and more quartzose, and the slate more micaceous as they approach, though an absolute line of juncture cannot always be recognised. It is not rare for the contact to be sharp, slate resting directly on granite, and this species of junction is most common in the non-mineralised rocks. The conduction of the granitic heat has been the cause of some curious and interesting transformation of strata ; thus, slates of fine texture have been changed to felspathic rock, sandy micaceous slate to mica slate, and even to a rock resembling gneiss, whilst some hard slates have become garnet bearing, and enclose patches of actynolite and axinite. Tourmaline schist is very common along the junction of the Hensbarrow, Tregoning, and other bosses, but it is difficult to say how much of these transformations may not have been effected by substitution since the consolidation of the granite.

Though the dykes of granitic matter called by Cornishmen elvans, have not mass sufficient to produce marked metamorphic changes on the slates which they traverse, yet in some places—where by some current, fresh igneous matter was perhaps passing—the elvan has pushed out veins into the contiguous strata, as at Poldory and Trevellas. At the United Mines there is an elvan which shows gradual transition to massive slate.

DIVISIONAL PLANES. The whole of the slate formations, together with the granite, greenstones, elvans, and even the veins themselves, are divided into innumerable rhomboidal and triagonal figures by a jointy

structure, the origin and age of which are as diverse as
the rocks they traverse. The manner in which these lines
of least resistance were first developed, has not yet been
satisfactorily shown, but as they are common to all
rocks,—whether subjected to plutonic action or not—and
acquire their fullest development when nearest the
surface, it may be due principally to a balance of the
inherent forces which constrain every mineral to assume
the crystalline shape appropriate to its elementary
composition, and to the exfoliation of the rock which
takes place when relieved from pressure, and exposed to
aqueous action. It is then evident that the Cambrian
acquired most of its jointy structure long before the
Silurian strata were deposited, and that an immensity of
time elapsed before the Devonian and Carboniferous
series were built up and similarly affected. Thus each
formation became penetrated by planes of division
peculiar to itself, besides some created subsequently,
which might be common to rocks of widely different ages.
It is in this way that the oldest sedimentary rocks of
Cornwall are usually those the most split up by planes of
separation. The metamorphic action which accompanied
the upheaval of the eruptive rock, though it obliterated
in its immediate neighbourhood the ancient joints, gave
after consolidation existence to a new series, which are
now seen to traverse indifferently granite, granite veins,
elvans, greenstones, and transition slates. The pro-
longation of joints from one rock to another, which is so
noticeable along the junction of granite and slate, cannot
therefore be considered as a proof of contemporaneous
origin. Where great pressure prevails, or when the rock is
of a nature little acted on by decomposing influences, the

joints though so prominent in the exfoliating rocks, are
in depth so close as to be often invisible, though their
existence is made evident when they are cleaved by the
quarrymen, to whom these planes of division are of the
utmost value and economy, the grain, as they term it,
enabling them to shape their blocks symetrically. The
divisional planes traverse the county at angles nearly
vertical, but are so much more prominent in some
localities, that they seem to affect a sort of grouping
arrangement. They are most persistent in two directions,
one oscillating principally between N.N.W. and N.,
whilst the other crosses them at right angles or nearly so;
a third group less conspicuous prevails capriciously in
some districts, and divides the rhomboids into rudely
triangular forms. There are other groups having
directions somewhat different, which appear to be affected
by the character of the rock which they traverse, as in
the serpentine of the Lizard, where they have a N. by
W. bearing. A most interesting feature associated with
the direction of the joints, is their general parallelism to
the east and west lodes, elvans, and caunters, and the
north and south cross courses. This connection is so
obvious, as to leave an impression, that the fissure
systems are but the final development of forces which
originated the divisional planes. Like veins, they are
not continuous, are subject to minor variations of division
and dip, and have a parallelism which displays itself in
groupings. Elvans and lodes, and even different portions
of the same, have sets of these parallel planes, which are
often filled out with quartz and tin ore.

In mineral districts the planes are more numerous and
definite, sometimes one series prevailing to the more or

less partial exclusion of the others, and they also run into each other both in strike and dip. They vary from an inch to several feet apart, have usually the thickness of a line, but are more open in favoured situations, and are then generally filled with clayey matter or quartz in slate, and with felspath and mica in granite rock. It is then apparent that the jointy structure associated with rocks and veins is of no particular age, but is a necessity to rocky matter, whether accumulated as strata, or segregated as metalliferous veinstuff. This inherent interstitial structure permits the circulation of currents of water through both sedimentary and plutonic rocks, as well as within the different vein systems.

CLEAVAGE. Neither jointy structure nor lamination of the slate beds have anything in common with cleavage planes, which though often imperceptible, and their existence only recognised by the facility with which the rocks split in their direction, pass through everything regardless of the undulations, foliations, or contortions of the disturbed strata. They pervade also the numerous fossils of the Tintagel neighbourhood, distorting them by elongation along the lines of lamination. The dip and direction of the cleavage is constant, and though sometimes nearly coincident with those of the bedding, crosses them at every conceivable angle. It is then evident that this property has been developed since the disturbance of the strata by plutonic force. Usually the slates effected have a grain so fine as to be indistinguishable, and some of them have such a perfectly fissile structure, as to be quarried for roofing slate ; this is especially the case in the Delabole district, where numerous quarries have been opened.

Notwithstanding that the cleavage permeates the whole
district, its fissility is, to a marked extent, dependant on
the rocks encountered ; for when the slates are not of a
slaty texture, but become siliceous, gritty, or of a loose
nature, the cleavage becomes less pronounced, and in
quartzose rock disappears entirely.   Again, when passing
from one rock to another—for instance greenstone to slate
—cleavage is not indifferent to the sudden change of texture, which gives rise to hesitating wavy lines.

On the outskirts of the Delabole district, where the
laminating force has been weaker, fine homogenous
grained slates may possess a fair amount of fissility when
the intercalated beds of sandstone may show but slight
traces.   The colour may be grey, green, or reddish, without the slates being necessarily different in constitution
or less obtuse to weathering.   The roofing slates are principally confined to the north of Cornwall, though some
are found in the vicinity of Saltash, and near Padstow.
The slates of the Delabole quarries dip westward at angles
a few degrees removed from verticality, and are divided
at intervals by east and west joints.   The slates produced
are of good quality, and, besides supplying the county of
Cornwall and Devon, are largely exported.

CHAPTER II.

ELVAN COURSES. Having touched on the jointy and laminated structure inseparable from rocks that have been subjected to pressure, when brought near the surface by upheaval and denudation, the description of the many diverse systems of veins subsequently developed follows naturally. To one accustomed to wander in the underground workings of the mines, the priority of the porphyritic dykes, which are known to all classes in Cornwall by the name of elvans, is manifested in various ways. That the range of granite hills through the two western counties are all connected by ridges now concealed under clay slate of varying but comparatively of moderate thickness, is made evident by a glance at the Geological Map, where the elvan dykes may be seen extending with scarcely a break of continuity from one eruptive boss to the other. Their formation during the consolidation of the superior crust of the granite, admits of small doubt, and though they may be considered geologically of the same age, the absolute distance in time between the genesis of the different groups must have been, humanly speaking, infinite. Their relative age may often be distinguished, the earlier elvans being recognized by their more granitic structure, and by their frequently acquiring a greater width as they approach the granitic nucleus.

The porphyritic elvans, which are much the most
numerous, fill fissures made probably through a much
thicker crust, at a period when the superincumbent strata
were less heated and allowed a quicker cóoling, so that
the crystallization of the felspathic matter was more or
less impeded ; consequently, these dykes are continuous
through granite and slate, and can be traced with facility
near the junction of those rocks, but with more difficulty
as the interior of the erupted rock is approached.   Elvan
has the same chemical composition as the granite from which
it proceeds, but its aggregation and structure, though probably identical at the bottom of the fissure, becomes less
and less granitic *pari passu* with remoteness from the
nucleus. The size of the dyke and the quantity of molten
matter poured through it, effects materially the rate of
refrigeration, and consequent constitution of the rock
consolidated. The effects of cooling are most conspicuous
at the sides of large elvan courses, or throughout them
when their size is diminutive. Thus they change quickly
in breadth but more slowly in length, for, however granitic they may be at first, they become highly porphyritic
on entering sedimentary strata, though, during a course of
some miles through them, they gradually lose this character, until at last, assuming a fine granular aggregation
they have much the appearance of sandstones, from which
however they are easily distinguished by their colour and
characteristic jointy structure. They seem to effect a
grouping arrangement, which may perhaps be due to
their occurring on hidden ridges of granite along which
the subterrene pressure has been more strongly directed.
However this may be, the greatest development of these
groups of elvan dykes is to be found running from one

granite boss to another, or taking a direction roughly parallel to the line of general upheaval. There are some localities which seem to confirm this idea, one for instance in Gwennap, where a network of elvans has the appearance of forming the spine of a spur of eruptive rock not far beneath. The Cornish elvans occur in the same manner as trap dykes, from which they differ chiefly in mineral aggregation, but are more plutonic in character. They fissure granite and slate indifferently, and run for many miles intersecting the beds at all angles, but preferably follow the lines of divisional planes, to which circumstance Mr. W. J. Henwood attributes their usually smooth and well defined walls. They are easily found by the numerous quarries opened along the outcrops, and are conspicuous underground by the crystals which they enclose, and by the contrast of colour and structure with the rocks in juxtaposition. In the granite their distinctness is not so apparent, but they nevertheless show a porphyritic character for long distances, and in one instance very nearly crosses the widest part of the Carnmenellis boss. Elvans rarely preserve straight courses either in strike or dip, but seize on the places of least resistance, and it is owing to this that they have a general wavy course that becomes abrupt when meeting with obstinate strata. Though these dykes keep as a rule a firm decided course, even those the most prominent are liable to split, a noticeable example of this occurs at Cayle near Hayle, where the Marazion elvan divides, one portion continuing to Pool village one side of a course of irestone—which may have given rise to the split—whilst the other to the north of it becomes the Roskear elvan.

If the fissures now filled with granitic matter, were

formed by fractures in the comparitively thin crust first
consolidated over the molten rock, these elvans must of
necessity become patchy and finally cease altogether in
depth; and there are symptoms at the bottom of the
deepest mines in the Dolcoath district, that their termi-
nation will be found at a moderate depth. It is seldom
that the laminæ of the slates assimilate in parallelism to
the dykes of elvan, but where they do so, fragments of
the rock traversed are sometimes found near the walls
along which the current seems to have carried them. An
instructive example is afforded by the Pentewan elvan
course, where, although they are very numerous at the
sides, none of the shattered pieces ever reach the central
portion.

The general directions of elvan courses varies but little
from those common to the divisional planes. The Land's
End tract seems singularly destitute of well characterised
examples, the patches observed usually appearing to
belong—like many of those of Hensbarrow granite—to a
period anterior to the formation of true elvans. In the
neighbourhood of Penzance commence the most considerable elvan courses in Cornwall; they have a remarkable
diversity of bearing near Marazion, where one runs N.E.
through Carn Brea district, while another extends
eastwards to Godolphin Hill, and a third takes a south-
easterly direction to Trewavas cliffs near the Looe Pool.
A most interesting group of elvans cross the county from
the Lizard through the East Wheal Rose Mines to
Newquay, and their diverse aggregation of minerals seem
to proclaim them of different epochs. Many of them
appear to curve away from the Gwennap dykes, to which
they bear a resemblance, others, such as the Trelissick

elvan, may possibly owe their origin to convulsions which heralded the approach of the granitic upheaval. The important groups of elvan courses that ushered in the formation of the St. Austell vein system, possesses a W.N.W. bearing, which is that of no other group in the county. The numerous elvans between Saint Columb and St. Wenn, indicate that the granite may be existant at no great depth, and their continuity and parallelism is indeed worthy of especial notice. These, like those associated with the granite hills of Bodmin and Dartmoor, run persistently from east to west. There is an extraordinary white elvan, composed of felspar and quartz, that runs from the neighbourhood of Camelford, several miles to Padstow, unaccompanied by a pronounced system of veins.

The dip of elvans is more regular than that of lodes, and though some of them have a flat underlie, yet very frequently their angles of inclination are less removed from the perpendicular. A great number incline at angles ranging between 50° and 60°, or from $3\frac{1}{2}$ to 5 feet in a fathom. Usually, and especially in the Camborne and Redruth neighbourhood, they incline from the granite.

The breadth of elvans are subject to such frequent and great variation, as to induce considerable suspicion that the hanging wall has slipped downwards during cooling. The width ranges from a foot to sixty fathoms, but the general size lies between two and twelve fathoms. They increase in average width from Marazion, where they are from one to eleven fathoms wide, to Gwennap, where their breadth is from three to thirty fathoms—but are again much smaller in the St. Austell group, in which district they vary from two to eight fathoms. The excessive variation in the width of even the same dyke is well exemplified at Wheal Unity in Gwennap, where the elvan—which was extensively excavated for the tin ore it contained—was in some places only six feet wide, whilst at others it swelled out to sixty fathoms.

The structure of elvan courses throughout all the metalliferous slates, bears a distinct relation to their remoteness from the parent rock, and the consequent more rapid cooling of the injected granitic matter. Their ultimate chemical composition may be everywhere much the same as the granite itself, but the aggregation of the minerals to which they owe their pronounced porphyritic character is dependant somewhat on the size of the dykes, but chiefly on the distance they have flowed through unheated strata. This is abundantly evident when the dull, hard, fine grained, and sometimes laminoid borders in contact with the walls of clayslate, are compared to the middle of the elvan, where the numerous brilliant crystals, sprinkled through a coloured granular base, produce a vivid contrast. This selvage, if it may be so called, is very prominent when the slates are of a loose schistose character, but is less marked when the walls are compact and crystalline. In the granite, the elvans having had their source from the mass below, are very similar in appearance and aggregation, but may be traced by their general finer grain, and by the compacter portions against the walls, whose texture resists weathering and remains hard even when the interior has been softened.

Some elvans, for instance those of Cubert, Padstow, and Blisland, have a very granitic base, but are rendered

porphyritic by disseminated crystals of felspar, which are
more numerous adjacent the sides. Throughout the
schistose strata of the county, the elvans have usually a
fine grained basis of felspar and quartz, though in
Gwennap it is more than ordinarily felspathic and compact,
whilst in many localities it is confusedly crystalline. This
base is rendered more or less porphyritic, by crystals of
felspar often large, and by translucent crystals of quartz
—which in the Gwennap group of elvans, assume the form
of sharply defined dodecahedrons—interspersed with less
numerous spiculae of schorl, sometimes pinite, rarely chlorite or hornblende, but nearly always some mica, whilst
in the Trelissick elvan course a mineral like bronzite
prevails to the exclusion of other crystals. Remote from
granite, the felspar crystals gradually disappear, and
after them the quartz crystals, until the elvan has an
even texture and granular appearance, like yellow sandstone, though still recognisable by the jointy structure it
owes to lateral cooling. This change of character is so
distinct, that one is tempted to judge the relative depth
of the underlying granite, by the intensity of the porphyritic character exhibited by the elvan. The diffused
crystals are sometimes so numerous and pronounced,
that they give a distinctive colouring to the mass ; thus
the black elvan of Chacewater owes its dark tint to the
unusual development of schorl and chlorite, the blue
elvan of Wheal Ann to hornblende and chlorite, and the
white dyke near Camelford to the absence of these
minerals, and the prevalence of white felspar.

The colour of the compact granular base is generally of
a light yellowish brown, but changing capriciously and
without apparent cause, it may be green and roseate, or

grey and drab, within a few feet, or even each side of a
line. The more felspathic an elvan, the more prone it is
to soften when exposed to the action of water, but this
decomposition is erratic; for the rock may be hard at
one place, whilst in another not far away it may be
worked for fire clay. In porphyritic elvans the felspar
has sometimes disappeared, and the cavities are
beautified by acicular crystals of various minerals
growing from the sides, or are completely filled with tin
oxide. More rarely—as near Terras—the base suffers
decomposition, and the crystals remain intact. Very
seldom, as on the cliff near Wheal Golding, an incipient
concretionary arrangement has become developed at the
surface of some elvans. As a rule, concentric lamination
is not natural to Cornish elvans, which affect a jointy
structure that divides the dyke into rude triangular
masses. The joints when open are filled with ochreous
matter, often with tin ore, and more rarely with
native copper. Although elvan courses have little in
common with lodes, yet they are often so called by
miners, on account of their enclosing reticulated veins of
tin ore when in slate, which mineral is also often disseminated in the elvans traversing granite; accidentally
other minerals are found disseminated in the rock,
amongst which, crystals of mundic, nodules of graphite,
and galena may be mentioned.

In Cornwall elvans have such a well recognised
influence on the segregation of ores, that no group of
productive veins in the clayslate has ever been discovered
unaccompanied by dykes of porphyry or courses of greenstone, and even in the granite, their influence is clearly
discernable. Tin ore is associated in the closest intimacy

with granitic rock, and is seldom, if ever, found far away
from it. Lodes when intersecting a compact felspathic
elvan, or following one of its walls, are, almost without
exception, found to contain large courses of tin or copper
ores, and sometimes the latter grows richer in the elvan
than in the granite or slate ; but should the porphyry be
jointy, the lode may, owing to its broken condition, be
unproductive. Elvans which permit free aqueous circulation, have had, especially in the killas, an important
influence on the enrichment of the "bunches" of mineral
in connection with them, but those hard, fine grained,
quartzose varieties, in mining parlance known as " dry
elvans," assist but little the deposition of metallic ores.
Elvan courses are not heaved to the same extent as the
lodes, with which they are so closely intermingled, and
this may perhaps be on account of their width, hard
crystalline and unyielding nature, and to their greater
verticality.

There is another kind of dyke, called, by Camborne
miners, Irestone which, though less plutonic and but
slightly porphyritic, has just as favourable an influence
on lodes as true elvans. They consist principally of
chlorite, quartz, and hornblende, are of remarkable compactness and induration, and are found running for long
distances with an average width inferior to that of elvans.
They seem older than the elvans, though the latter rarely
intersect them, prefering to take a parallel course. They
have the same bearing as lodes and elvans, and are at
Roskear and Tincroft, about twenty fathoms wide. An
eminent mining authority in the Camborne district has
remarked, that in the killas, no mineral of value has been
discovered apart from courses of elvan and greenstone.

VEIN SYSTEMS. The instability of the terres-
trial surface, which a study of the rocks of Cornwall
reveals, gives a considerable shock to one's ideas of the
solidity of the earth ; and the ocean, the archetype of
changefulness, is found to have had an eternal existence
compared to the time during which a series of strata became deposited, and by denudation restored to the ocean.
Doubtless the granite was denuded on emerging from the
ocean, but its true crystalline texture proclaims the vast
depth of sedimentary strata with which it was covered
at the period of consolidation. Though it is difficult,
with our imperfect knowledge of what may happen some
miles below the surface, to designate a theory which
would be capable of explaining the origin of the diverse
fissure systems, which have, time and again, ruptured
the clay slates, we may conjecture that they were formed
immediately after the elvan courses, when the granitic
crust had acquired an increased thickness. The fissures
were possibly the result of earthquake shocks, widened
by shrinkage and general upheaval of the slates ; but in
whatever way they may have originated, they permitted
freer access of water to the heated rocks, and introduced
conditions eminently conducive to the segregation of both
earthy and metallic minerals, and to their deposition in
favourable cavities at a distance appropriate to their
several affinities. The immensity of time demanded for
consolidation, allowed ample opportunity for the formation of fissure groups, for the dislocations and faultings
which developed the hollows, and for the gradual filling of
these by mechanical or thermo-electric deposition. The
minerals would appear to have been deposited according
to the resistance they have offered to heat and pressure ;

Junction of Killas and Granite.

S⊤ MICHAEL'S MOUNT

Publ.d by H. Besley, Directory Office, Sandford St. Exeter.

in this way can be perhaps partially elucidated the association of the richest deposits of tin ore with depth, and the occurrence of galena at a distance from the intensest thermal action, whilst cupreous deposits seem to occupy an intermediate position. The association of tin and copper in the same vein. may be partially explained by the re-opening of the tin lode, and subsequent deposition of cupreous ores between it and the dislocated hanging wall. The two parts of the lode are usually separated by a " parting " of clay or quartz. The whole county having been subjected to fissure systems produced at numerous epochs, during which the existing veins were many times opened, faulted, and the resulting cavations again filled with lode stuff; the elucidation of all the phenomena of heaves, and of the occurrence of metallic mineral, is attended by insuperable difficulties.. Notwithstanding that in some instances, the so-called heaves may have been small in amount, it is certain that the vertical or very slightly oblique downward movements caused by the cross courses. is of much greater magnitude than the earlier geologists of Cornwall conceived. Though the surface is now so level, a study of the heaves in the St. Agnes and Perran districts, discloses an aggregate downthrow to the south, caused by the east and west lodes of more than a thousand feet; while the cumulative slip east, due to the cross courses, has been estimated by Mr. J. H. Collins, F.G.S., at many thousands of feet. In the face of such enormous dislocations, together with the probability that slips of strata have taken place between the cross veins, unaccompanied by much movement of the latter, the recognition of the ends of a heaved lode must be of extreme rarity.

E

The definition of true fissure veins is not altogether so easy as might appear; broadly stated, they may be said to be clefts of varying width, extending from the surface to such deeper portions of the earth's crust, as have a temperature and pressure sufficient to cause such hydro-thermal circulation through them, as will lead to their being filled with minerals of a character different from the strata which they have cut through. It would be well to mention here that beside the true fissures which traverse granite, slate, and elvan indifferently; there are numerous non-persistent veins, which, as they are confined to rocks of a definite epoch, are usually spoken of as contemporaneous. These veins are not metalliferous, but contain only earthy minerals; they are found all over the county and are filled with felspar, mica, schörl, quartz, chlorite, actynolite, thallite, axinite, prehnite, garnet, serpentine, asbestos, agate, jasper, opal, &c.

RELATIVE AGE OF VEINS.

Few places in the world can compare with Cornwall in the number and diversity of veins, and the anomalous phenomena which have accompanied their formation during a long continued succession of disruptions. For the sake of perspicuity the lodes and cross courses have been arranged in groups, and a relative age assigned them dependant on the intersection of one vein by another; and, though there may be a few cases where a vein crossed by another may yet be the youngest, there is no better test of relative age, than the matter of one vein passing through that of another. Mr. Carne, at an early period

in the geologic history of the county, devoted himself to
the study of its vein systems and their priority of formation. The attempt to separate such a complicated series
of fissures, with an approach to correctness, into definite
systems possessing the same age, would be impracticable,
on account of the numerous movements which older veins
unquestionably suffered, when reiterated convulsions gave
rise to new fissures. As however some sort of provisional
classification, though to some extent erroneous, is indispensable in order to describe intelligibly the occurrence
of lodes and cross courses, they may—if we trust to the
evidence of intersections—be distributed in seven groups
as follows, viz. :

    1.   Older stanniferous lodes.

    2.   Newer stanniferous lodes.

    3.   Older east and west cupreous lodes.

    4.   Caunter lodes.

    5.   Cross courses and Flucans.

    6.   Recent lodes.

    7.   Slides.

Although the cross courses and flucans are placed in
the same group as if they were of similar age, there are
many examples of their crossing and heaving each other ;
and had their intersections been exposed to the same
amount of observation as those of lodes, it is possible
that considerable diversity of relative age would be apparent in them.

The development of parallel groups of productive
veins, is influenced to a marked extent by the general
character and texture of the geological formations disturbed by the eruptive rock. If, for instance, the Cornish
strata had been of homogenous character, or composed

of ancient rocks already highly metamorphosed, instead
of alternations of clay slates and sandstones of diverse
grain, whose separate homogeneity, induced a wavy direction of the fissures due to the refraction of the subterrene shocks; the vast metallic wealth stored in the
cavations resulting from the downthrow of the hanging
walls, might have had no existence, or at least the
deposits would have been comparatively insignificant. It
is undeniable that veins are the largest and most persistent in compact and thick lamellar strata of diverse grain
and density, particularly when strata incline gently outward from the granite, with a strike parallel to its flanks
and to the line of general upheaval. Supposing indeed
that the whole of the clay slates were nearly of a similar
texture, density, and composition, the fissures finding
nothing to refract them, would make an almost plane
fracture, whose faulting would unfold hollows of insufficient importance to allow of that energetic hydro-thermal
circulation, to which deposits of the useful metals are
generally believed to owe their origin.

It has been remarked by the earlier investigators of
the fissure systems of Cornwall, that the cumulative
thickness of all the lodes known and undiscovered, would
amount to an incredible proportion of the rocks; but it
must be remembered that the fractures were formed suc-
cessively by long continued upheaval, during which the
clay slates lost much of their bulk by shrinkage, and the
fissures acquired access of width by abrasion, aided by
intense thermal action, which dissolved much of the wall
matter that subsequently subsided in those portions of
the veins, where a lower temperature prevailed. Though
some groups of veins (especially those to the south of St.

Austell, and St. Just) have a widely different bearing, yet the general strike of the lodes—which coincides so singularly with the great lines of granitic eruption which run from the Scilly Islands to Exeter—seems to demonstrate that though intermittently formed, they, together with the cross veins, are nearly all the result of the same general epoch of upheaval. The fissuring extends throughout the county, but lodes become more numerous as the eruptive rock is approached, and are best developed in the slates which repose immediately on the granite. The influence that the mechanical condition and dip of the strata exercises on the development of veins, is made manifest by their absence where shelfy strata predominate, or where the strike of the beds is not concurrent with the contour of the granite; and, by their number and size, where the strata possessing the necessary favourable characters prevail. Viewed comprehensively, both lodes and cross courses thin out after running for some distance, but the fissuring continues in the same direction. It is not at all certain that the same vein extends without a break more than a couple of miles in length, but it is generally believed that the County Cross-course has fractured the rock from sea to sea. Although the lodes extend so considerably in length and depth, and are so important in regard to their metallic produce, yet they would resemble, even at a large scale, only a sheet of paper in size.

Every considerable system of fractures has its appropriate set of cross veins which have a relatively rectangular direction. The slate beds resting on the granite adjacent the junction, though so completely shattered by fissures of diverse direction, have been subjected to less downward movement of the hanging walls of lodes and

cross courses than those at a distance. The amount of
dislocation is in evident relation with remoteness, for
while the apparent faulting of veins near the granite is
seldom extreme, at a few thousand feet the downthrows
become very considerable. Observation has been little
directed to the very difficult study of downthrows, but in
St. Agnes—where the frequent intersection of two series
of lodes with opposite dips, has revealed the important
movements to which veins are liable—the evidence seems
to suggest that the granitic cones may have received some
part of their elevation because they were forced up along
these fissures by the upheaval which produced them. In
some districts where strata of contrasted character prevail, the extent of faulting is made visible, but in Cornwall the clay slates are so similar, that the slip of the
hanging wall can seldom be measured. The faulting,
caused both by veins and cross veins, generally appear
cumulative in a direction from the granite; and the
downthrow, especially of the crossveins, is not often the
result of one fissure, but is made up of a number of
parallel veins, which often compose what is called a cross
course. Even the east and west lodes loose their mineralized character when they are distant from the granite,
and become faults filled only with clay or quartz.

The grouping of joints and of elvans has not escaped
observation, and the lodes and cross veins which follow
nearly their direction affect a similar parallelism. There
need be no hesitation in believing that these groups are
due to distinct shocks affecting a certain assemblage of
slates by which they are often limited. When such a
collection of contemporaneous lodes occur in favourable
strata, they are—being subject to the same conditions—

commonly productive across a zone that coincides approximately with the bearing of the associated cross courses ;
an occurrence which the miner expresses by " ore against
ore." At whatever time, during the elaboration of the
metalliferous deposits as now seen, veins may have been
formed and filled; subsequent shocks not only formed
new groups but opened again the older fissures. The
lines of fracture did not necessarily take the plane
of the wall, but frequently crossed the vein irregularly
giving rise to " partings " in the gangue. In this way the
size of lodes was considerably augmented ; and this mode
of increase is remarkably developed in Crowan and
Gwinear, where the " comby " quartzose lodes supply the
most conclusive evidence of repeated enlargment. In
whatever way the downthrows may have been produced
the result in every case is to lower the hanging wall,
which sliding down on the waving unequalities accompanying every fissure, occasioned a succession of cavations
that became filled with the valuable metalliferous products, alternating with barren " bars of ground " at those
points where, the two walls being in close contact, the
deposition of mineral was obstructed. The deposits in
lodes, whether of earthy or metallic minerals, are due to
the choking up of the ancient water channels by precipitation.

Lodes have been spoken of as very absolute in their
course, but though they may be ever so persistent, the
miner could "a tale unfold " of the troubles experienced
in following veins through the "nips" where the opposing
walls meet and leave only a line as indistinct as a divisional plane by which to trace them, when the lode is said
to be " lost." This difficulty is greatly aggravated by the

shattered nature of the lode, which is rarely if ever
simple, but splits off into parallel **branches** that may
rejoin the same or another lode, or may dissipate itself in
the slate, and sometimes an insignificant string may leave
the lode and become the productive part. That most of
these branches and splits are contemporaneous, may be
inferred from observing those portions of rock known as
" horses," which are often included in large lodes,
especially where strata of diverse appearance are crossed,
when the surface of parting will be discovered in the opposing walls. Sometimes lodes split and diverge considerably from each other, as in the Marazion and St. Just
districts, and occasionally in connection with cross veins,
as at Great Devon Consols and Polgooth. There are
numberless examples of " blind leads," or veins that do
not reach the surface, but they may be placed in the same
category as "**droppers**," which fall away from the
lodes. Branches and veinlets are known to the miners
as " droppers " when they leave a lode ; when they
dissipate themselves in the " country " they are usually
considered to "bleed" the lode, but "**feeders**," the
lateral branches which fall into a lode, are believed to
have an enriching influence when the angle of incidence
is small. Joints are even supposed to have a similar
effect when they fall into the lode. By these innumerable fissures, the mining districts have been shattered
into huge lentiles of rock, more or less cemented by the
mineral deposited, which owing to the jointy structure,
are subdivided into small fragments of triagonal shape.

The **strike** of the vein systems of Cornwall is in
remarkable accordance with that of the true elvan groups,
even when these are discordant—as at St. Austell and

Marazion—with the general trend of the granitic upheaval.
The lodes subordinate to each eruptive boss, have a direction appropriate to it, though this *per se* seems to have
exercised little modifying power on their character, as
caunter lodes bear no distinctive impress apart from that
of "bearing." The important groups of meridional elvan
courses that cross the county at Truro, separate vein
systems of very different strike; for while those to the
east have an oriental bearing, others between Truro and
Penzance run with great persistency to the north-east.
In the district of St. Austell the average bearing is north-
east by east, in that of St. Just north-west by north, the
latter bearing agreeing very nearly with that of the
caunters. In the district lying around Godolphin Hill,
the lodes assume indifferently all three directions. In
every district the lodes are subject to warpings, that have
a very decided influence on their metalliferous contents,
and observation has brought out the fact, that the bearing of rich parts of lode are as a general rule nearest to the
average direction of the enclosing vein, especially when
it coincides with the general line of upheaval.

The great bulk of the copper and tin ores raised, has
been obtained—according to Capt. Charles Thomas—from
deposits that affect a magnetic bearing (1860) ranging
between five degrees north, and twenty-five south of east.
To give the average bearing of the lodes in each district
would serve no purpose, and might indeed lead to
erroneous ideas, because the average bearing of one series
of veins, would be confounded with that of another whose
formation in time would be different. Lodes of diverse
bearing are chiefly confined to the extreme west of Cornwall, where two, and often three sets of lodes are preva-

lent ; eastward it is seldom that the lodes are intersected
by caunters. It is not probable that lodes ever cross large
tracts of granite, because they are naturally restricted to
that portion of the eruptive rock which was consolidated
at the time of their formation ; but that they cut through
smaller bosses, is evinced by one or more of the Polladras
tin veins, which, fissuring Godolphin Hill are worked on
the southern slope at Wheal Grey. This phenomenon
is interesting as tending to confirm the theory, that the
formation of Cornish veins is due to forces acting on a
comprehensive scale, far outside any influence exerted by
isolated hills of granite.

INCLINATION OF VEINS.

The dip of lodes, owing to varying direction of the plutonic action during upheaval and consolidation, is very
irregular, yet there is some coincidence between the inclination of the veins and that of the lines of pressure
caused by the elevation of the granitic masses.   Comprehensively speaking, it may be said that in the central
portion of granitic tracts, the lodes have scarcely any distinctive hade, that on their flanks, in the majority of
cases, the dip is towards the granite, whilst the rest incline from it, and these latter increase in number and
verticality, until in the killas remote from the intruding
mass, they become again nearly perpendicular.   There is
a considerable variation in the amount of dip, the few
called "flat lodes" falling only thirty degrees from horizontal, but the mean inclination of lodes from Dartmoor
to the Land's End, is—according to Mr. W. J. Henwood

—about seventy degrees, and the same authority states, that nearly two-thirds of the veins hade to the north. However much the inclination of any portion of a lode may vary, it returns towards the mean plane, and the reverse of the dip is of rare occurrence, except in the case of a perpendicular lode. The different angles which a vein pursues in depth, is mainly due to the refraction of the initial blow by alternating strata of diverse texture or density. The range of angular inflexion in the same lode, is from 10° to 25°, and in exceptional examples even 35°. Thus, wavy planes are produced, and any movement of the strata occasions a fall of the hanging wall ; this will leave hollows that must of necessity be the most vertical portions of the lodes, and of course those which must be the receptacles of mineral deposits. Therefore the richest deposits are found enclosed in the compactest rock, in those portions of the lode nearest to verticality, and a change to a flatter dip would naturally be indicative of approaching poverty. Flat lodes yield copper or tin ores of inferior quality, but they nevertheless are submissive to the above rule. Though veins of little inclination may be comparatively poor, it is impracticable to fix a hard and fast line for the most productive dip.

Size of Veins. From what has been said above, it may readily be recognised that width is the most uncertain property of a lode. Its variability is such, that from a line with difficulty distinguishable, it may swell out to a valuable deposit ; this does not happen suddenly, but the breadth increases gradually, the metalliferous portions proclaiming their advent by the admixture of stones of pyrites or blende, and of the metal which is the chief product of the lode. The greatest width of a Cornish

lode is sixty feet, but such magnitude is rare. Beside the
difference in size produced by faulting, in passing from
one rock to another the softer portions of the walls of a
vein may be crushed together by pressure. Fissures
when first formed were mere lines and their enlargment
was due principally to the faulting occasioned by reiterated shocks. Lodes in granite are somewhat smaller
than those in clay slate, the average width in the former
(Mr. W. J. Henwood) being 38, and in the latter 45
inches. In mining parlance, a lode is a rock from which
metalliferous ores are extracted, and in this sense many
lodes have a much greater width than that of the associated
fissures. This is owing to the brecciated condition of the
" walls," and to their impregnation by the percolation of
mineral waters, which not only fill the interstices, but
may dissolve the rock and substitute in its place the
mineral in solution. The size of lodes appears to have a
close connection with the priority of their formation, if
we take the mean width as deduced from the careful and
laborious measurements of Mr. W. J. Henwood, viz. :

Tin and copper lodes	...	feet	4·70
Tin lodes	... ...	,,	3·06
Copper lodes	... ...	,,	2·93
Lead lodes	.. ...	,,	2·00

This is in accordance with the theory of the re-opening
and faulting of veins, because the oldest would naturally
be subjected to the greatest number of movements. The
extra size of the tin and copper lodes may be attributable
to re-opening, or to the junction in strike and dip
which would cause them to run on together.

BACKS OF LODES.

We can only conjecture whether or not the veins systems of Cornwall at the time of their formation ever reached the surface; but as the large granitic veins, by their arenaceous structure and dissipation amongst the slaty beds when remote from the eruptive rock, seem to be confined within a limited zone; it is logical to infer that vein fractures would not have a more penetrative power. As a consequence veins may perhaps exist whose "backs" do not reach the surface. With few exceptions the "backs," or basset edges of the lodes near the surface, to a depth that does not often reach much below the level of the vallies, have been much transformed by oxidation; a result which the miner calls "bleeding of the lode." The extent of this change depends on the depth to which rains containing carbonic acid and oxygen can penetrate. Owing to this chemical reaction the gangues have become " Gossans " to a limited depth irrespective of the surface undulations, and in exceptional cases have been seen 50 fathoms from " grass." The alteration is perhaps most profound in granite, because of its superior elevation, and the facility of circulation afforded by cross courses. It must not be imagined that gossan is always concurrent with the " backs," on the contrary, it is, except in some very favoured districts, oftener absent. Its presence is dependant first, on the character of the contents of the vein which is not always metalliferous, and again on the proximity of the enclosing walls, whose opposing undulations create such a difference, that though a lode may have a fair average width, it is often so squeezed as to be untraceable at the surface.

The composition of gossans is of course entirely subordinate to that possessed by the original matter, they
are therefore chiefly quartzose and ferriferous. The very
hard, siliceous, and non-metalliferous portions of the
backs, are obtuse to oxidation and have suffered little
change. As a similar character distinguishes true tin
lodes, their backs show gossan only imperfectly, but sometimes have been worked for the tin disseminated through
them. Those on the backs of copper lodes are irony, less
quartzose, have a decided vesicular structure, and are
more friable than tin gossans. The minerals contained
in the drusy quartz nearest the surface, are earthy brown
iron, black copper ores, green and blue carbonates of
copper, blende, galena, and occasionally spangles of
native copper and cupreous arseniates ; indeed, the greater
number of curious and rare minerals have been obtained
from the shallow portions of lodes, and are seldom found
in connection with productive bodies of ores. The transition from the gossans to the courses of ore which they
conceal is not abrupt, but the ferruginous and quartzy
substances are gradually replaced by mundic, yellow copper ores, zinc blende, and other sulphides, until no traces
of oxidised metals remaining, the productive deposits are
reached. In not a few instances a gossan may be the
back of a deposit of sulphide of iron or arsenic, but as
mundic " rides a good horse " a cupreous deposit may be
expected at an increased depth. Notwithstanding that
gossans may occasionally be deceptive, one that the miner
would recognize as a true ferruginous gossan, has never
failed to cover rich deposits either in Devon or Cornwall.
The finest lodes around the Carnbrea Hills, and the celebrated lode of Devon Great Consols were all concealed

under masses of splendid gossan. Miners have predilections for gossans of a certain colour and hardness, which are not always the same in different districts. The usual colour of a good gossan should be brownish red of various tints, but a poor gossan is generally composed of a milky white quartz often mingled with slaty rock, more or less destitute of ferruginous hues. As a rule the backs of tin lodes are much harder than those of copper or lead lodes. Gossans though not of themselves valuable, often become so through admixture of gold and silver, which is sometimes present in quantity sufficient to be successfully worked. The want of gossan does not necessarily stamp a lode as unproductive, though it is considered by miners as an unfavourable indication.

The mechanical **Structure of Lodes** is simple, and their planes, joints, and interstices, are evidently due to the same inherent causes which have produced those seen in any consolidated rock. They were developed during the formation of the vein, but subsequently to the deposition of any separate portion of it. The quartzose parts of veins ore often " comby "—and singularly so in the parishes of Gwinear and Crowan, being made up of a number of thin lentiles of quartz with definite walls, which owing to the movement of the walls at diverse epochs, are deficient in continuity. In poor lodes, these lentiles are composed of white crystalline quartz whose opposing serrated surfaces create drusy cavities which are sometimes of considerable size. In more mineralized strata, the lentiform masses are extensive and compact, and are so similar to the " country " that even the miner may be deceived, and follow a plane that is not the true wall of the lode, to discover which " crosscuts " are

driven across the walls. The general displacement to
which the encasing rocks have been subjected even in recent times, is made apparent by the planes of moist plastic
clay which are found along the wall of the lode, and between partings of the same lode. Even a compact mass
of " lodestuff " often confirms by a conchoidal structure
with polished striated faces called " slickensides," the
movements to which the filling matter itself is liable.
The substance of the lodes is also divided by joints which
are often a continuation of those from the enclosing rock,
and at times a horizontal lamellar structure exists across
the vein.

Walls of Lodes. This is a subject which has been
much discussed by people, who, from the similarity often
existing between the walls and veinstuff, have supposed
that much of the contents of lodes may have been segregated from the " country " rock and some have even believed in their congenation; but, notwithstanding that
originally a minute fraction of the more soluble ores may
have been enclosed at the time of the deposition of the
strata, it is more in harmony with the conditions that
regulate the segregation of mineral matter, to suppose
that the metalliferous character of the adjacent rock, is
attributable to infiltration and substitution during the
filling of the veins themselves. That smoothness of wall
which is so important a characteristic in productive
courses of ore, may be the result of an initial fracture in
a compact lamellar rock, or to the fissure being concurrent
with the previously existing jointy structure. Where
these conditions are absent, the walls are liable to be
rough and even jagged, and not likely to enclose large
bunches of ore. The opposing sides of valuable lodes,

DIAGRAMS TO ILLUSTRATE GROWTH OF VEINS

Third descent of Wall   Second slip of Hanging Wall   First opening   Fissure birth

COMPACT   ROCKS

COMPACT   ROCKS

d   c   b   a

could never be adjusted, on account of the change of position, and the frictional abrasion produced by the descent of the hanging walls, whose protrusions must have crushed large fragments off the walls which fell either to the bottom of the hollows, or became a breccia in the side of the lode.

Walls have, accordingly, a very varied structure, and may possess a sharp line of division, a brecciated character, or may pass into the contiguous rocks by such insensible gradations, that it would be difficult to point out where the gangue terminates and the " country " begins. In granite where the vein matter derived from the contiguous country retains its appearance with greater obstinacy than in clay slate, the transition, though also gradual, is accomplished in less distance. In schörlaceous lodes the layers of schörl and felspar change so gradually, that no definite stratum can be safely selected as the true wall. Slabs of contiguous rock split off along a joint, become by the removal of the soluble parts and the infiltration of quartz or metallic minerals, true portions of a lode. The wall of a lode is often clearly defined, and especially so when the gangue is quartzose. In some lodes—particularly interesting in the parish of Gwinear—boulders, like concretionary lumps, were found in the walls, and were possibly the first filling of a fissure caused by friction, which, in the absence of mineral in solution was consolidated by earthy matter. But in numerous other instances in that and in other districts, this debris, cemented by spar and sulphuretted ores, or by tin oxide, formed the lode, and proved the presence of fertilizing solutions. The brecciated character of the fissure walls, is, as might naturally be anticipated, very prevalent in all districts,

F

whatever metal the lodes may produce. As the separated pieces are forced off the points where the hanging
wall crushes on to the foot wall, and the fragments fall
into the bottom of the hollows, this brecciated structure
is most conspicuous where the lodes become flattened.
They occur both in granite and slate, are generally termed
the " capels " of the lode, and in many instances they
pass into " country " without any plane of separation.
These capels much increase the size of the lodes, and
often contain a workable percentage of cassiterite. The
useful parts of lodes also become larger than the original
fissures, both in granite and slate, by the infiltration of
mineral matter into the interstices of a shaken portion of
the wall, or into the rock itself by substitution. This
mode of lode creation is exemplified in an interesting
manner in schorlaceous granite, where rich bunches of
ore often accumulate on one or both sides of a slender
veinlet.

In cross courses the walls are often still less discernable, on account of the number of fractures, and the
occurrence of lateral quartzose veins and strings, which
also make their appearance in the cross course. In the
granite the filling bears even more resemblance to the
encasing rock, and a true wall is frequently undiscover-
able. It is owing to this incorporation of the rocks in
juxtaposition, assisted by the freer percolation of pluvial
waters in them, that the mineral character of the strata
indicates by a peculiar softening change in them, the
proximity of a lode. The " Keenly ground " is of course
due to the presence of the fissures, the walls of which—
especially in the deeper veins—may have been subject to
a mild metamorphic action, which predisposed the rock

to change when subject to the percolation of waters containing carbonic or other acids in solution. In some districts, notably at St. Austell, the lodes stand out like stone walls, owing to the decomposition of the felspathic granite which they traversed. It should be added, that to imagine all metalliferous gangues to have originated in the same manner would be erroneous; on the contrary, the method of vein formation and the segregations of their contents differ according to the characters of the rocks disrupted, the way in which the dislocations have occurred, and to the geological age of the formation which the veins have fissured.

Lodestuff or Gangues. From what has previously been said, respecting the alternating hollows and pinches which the descent of the upper wall must of necessity occasion in an undulating fissure, it will be manifest that the matrices which enclose the metallic ores must have a distinct reference to the mineral character of the contiguous strata. The change in the rock traversed is therefore immediately followed by a corresponding alteration in the earthy, and frequently even in the metallic, filling of the lode. Though the filling of the hollows by vein stuff has been due to various causes acting at divers periods, silica is associated with the matrix in every kind of rock, and is usually the most prominent ingredient. In clay schists, the gangue is composed of quartz and slaty matter; and in granite, of quartz and decomposed granitic debris. A careful investigation of the contents of veins, gives striking evidence of the complication of causes which has determined their segregation, and of the mode in which the lodes acquired their size by successive movements and refillings. (*Vide* PLATE I).

As the percolation of water would naturally descend
through the interstices towards the heated zone lying
between the unconsolidated granite and the clay slates,
the fissures were probably filled at first with water at
high heat and pressure, and therefore in a suitable condition to hold in solution both earthy and metallic minerals, and the aqueous circulation set up by the cooling of the
superior stratum, would conduct them to fitting localities
in which they would be deposited by subsidence, or by
precipitation amongst rocks having the requisite thermo-
electrical affinities. In dessicated strata undergoing
metamorphic action, metallic minerals may have passed
through in a state of sublimation, and been deposited in
veins amid colder rocks. The perplexing manner in
which ores of different metals are found intermingled,
may be due to some change in the character and strength
of the ascending currents ; because, it is well known,
that a trifling alteration in the condition of a solvent will
retard or induce precipitation. Besides the slaty and
quartzose matter which go to make up the bulk of the
gangues, there are other earthy minerals met with but in
far smaller quantities, such as calcite in the northern part
of the county, fluor spar, &c.

In the granite districts the earthy minerals are
in different proportions, for though quartz is still the
prevailing mineral, felspar, schörl, and mica are more or
less prominent, which give the gangue a somewhat granitic
aspect. A faint distinction can be recognised between
the lode stuff of veins of different formative eras, the tin
veins having a more compact crystalline texture, whilst
the filling of the copper lodes is more open and often
separated from the walls by flucany selvages, that become

still more pronounced in the cross courses. Flucan receives its full development in the cross flucans and slides, which are admittedly among the youngest dislocations, and are exclusively composed of clayey substances. There is but little evidence of particular arrangement in these earthy gangues, for although there is often a "leader" which runs for some distance both in length and depth, it is perhaps more customary to find an irregularly mixed gangue.

The heterogenous character of the veinstuff is more than reflected by the eccentric mode in which metallic minerals are distributed in it. Unquestionably the most abundant are sulphides of iron and arsenic, better known to Cornishmen as mundic and mispickel; then comes blende or black jack, and wolfram. The most abundant of the useful ores is copper pyrites, tin ore, galena, and manganite. These ores occur in veins occasionally associated with others less common, such as barytes, antimony, silver, uranium, bismuth, and molybdenum. Tin is the only oxide existing in depth, the metals being nearly always mineralised by sulphur and arsenic. Though the mixture is irregular, the passage from earthy to metallic minerals is never abrupt, but a " course of ore " is gradually developed by the appearance in the matrix of what the miners call "stones of ore." The useful ores, with mundic and arsenic, occur in the lodes in the form of strings, bunches, and courses of ore, sometimes " clean " but generally contaminated with a changeable proportion of the earthy minerals above noticed, quartz being always predominant. This matrix containing no mineral of value is called " deads," and is either put to "stull," that is thrown in the " gunnies "

or cavites left by previous excavations—or hauled to the
surface and trammed over the " burrow tip." None of
the lodes produce ores throughout their length, but
chiefly in those hollows caused by the flexures of the
fallen hanging wall where the sides—especially the foot
wall—are defined and compact, with a bearing approaching
the mean direction of the lode. Therefore courses of ore
necessarily occupy the most perpendicular, and broadest
portions of the lode, and it follows that when these
productive channels dwindle and the lode becomes
pinched, a corresponding "bar" or alternation of barren
ground must be anticipated. As the courses of ores,
whether tin, copper, or lead, fill hollows which are usually
confined to the compactest alternations of the clayslates,
and as the latter slope away at gentle angles from the
granite, it would be expected that the rich parts of the
lodes would incline from the eruptive bosses. This is
found with very few exceptions to be the fact, and not
only in the slate but even the granite, the deposits affect
a dip, having a relative coincidence with the outward
inclination of the granitic grain, due to conditions
developed by its junction with the sedimentary strata.
There are exceptions to this peculiar, and to miners, very
important circumstance; for instance, where the clay
schists have a bearing nearly parallel to that of the lodes,
when faulting produces hollows with little obliquity, and
the deposits may then often approach a columnar outline;
and where the horizontality of the slates having remained
undisturbed, the courses of ore might have little dip in
any direction. It will thus be comprehended why, if one
of a group of veins in strata of alternating compactness
possesses rich courses of ore, the corresponding parts of the

parallel lodes may be expected to yield results equally good. The prevailing dip of the productive portions of a lode, or of a group of lodes, is an object of much interest to the pratical miner, as much capital may be expended in a wrong direction, when the dip of the "country" is not recognised; and therefore by careful observation of the diverse density of the strata, and their relations to the strike and dip of the veins, he strives to attain the requisite knowledge.

Lodes are called productive in the sense of paying, but they may be rich in ores for which there happens to be no demand. Formerly many sulphides, such as arsenic and blende were treated as "deads" that are now remunerating the miner for his labour. A good course of ore should be massive, compact, with an indefinite structure; but comby quartz, with a pronounced jointy and horizontal bedding having defined partings and drusy cavities, are unequivocal symbols of poverty. Veins which cross groups of productive lodes are almost without exception poor, nor are crooked lodes very favourable to deposition of mineral.

True courses of ore, which are generally persistent in length and dip, may be more easily found than other deposits in veins, many of which have had their genesis at subsequent periods, and which on that account have been designated by M. Moissonet "accessory" or "accidental riches." There can be no doubt that after the formation of much of the veinstuff, reactions of an interesting character happened. These abnormal changes are most clearly exemplified in the remarkable accumulation of ores in the carbonas, pipe veins, and amorphous unconnected masses of mineralised granite in the Land's

End district, and in the changes affected where schörl abounds. Even in the most recent epoch, the frequent occurence of pseudomorphs, clearly shows that the gangues and ores of lodes are exposed to incessant alterations; whilst there is reason to suspect that movements of the walls still take place, and may in the present, as they have in the proximate past, given rise to those eccentric bunches and strings of ore whose mode of segregation is so obscure. Carbonas appear to owe their presence in the granite, to the infiltration of metals in solution along some irregular vein or joint, near which the surrounding mass becomes impregnated by precipitation and substitution. The rich floors, bunches, and veins of tin ore in the schörlaceous granite of Hensbarrow, are due to precisely similar action. Important **Stockwerks** have been opened where these impregnations abound at Carclaze, Beam, Minear, and others in the granite; and at various places in the clayslates near Bodmin, from all of which much cassiterite has been obtained. Copper ores have also been raised from **Stockwerks** in the schistose rocks at the Bunny mine in St. Agnes.

CROSSCOURSES and FLUCANS.—These veins cross the county with a mean direction approximately the same as the faults which traverse the Palæozoic rocks of Wales and England. They are variously known amongst Cornish miners by names having reference to their contents or other peculiarities; thus they are termed in the vicinity of Camborne and Tavistock, crosscourses if they contain quartz, and flucans when they are filled with clay. In the west they are called cross gossans when enclosing ferrugiuous quartz, "trawns," and "guides" because they lead to productive lodes. In the Saint Austell district they are distinguished as iron lodes, because many of them have been worked for iron ores. So vague is the signification of crosscourses, that in a lead district the east and west quartzose veins, which intersect the productive lodes are also known under that name. Their unmineralised character has afforded no incentive to exploration, and they have on that account not attracted the same investigation as metalliferous veins; consequently, the dislocations to which they may have been subjected, are inadequately appreciated. For this reason an estimation of the age of cross veins, in relation to that of the diverse origin of lodes cannot be made. Although it is a generally accepted fact that meridional veins are the result of the last upheaval shocks, it is not improbable, that many of them were produced during the re-opening of the older, and formation of some of the younger veins; convincing proof of this is wanting, owing to the downthrows of the cross veins, and the tendency of the last faulting to obliterate the effects of previous slips. But in some instances near the granite—where the fall of the hanging wall is always less in amount—evidence of contemporaneity is palpable. In many cases the filling matter of the crosscourse between the heaves, partakes of the mineral character of the lode, and even in some examples, the metallic minerals are carried along the whole extent of the heave, and the course of ore made continuous through lode and crosscourse. At Wheal Friendship and at Ting Tang mine, this occurred in a most marked and highly interesting manner; a course of vitreous copper ore, two feet wide, extended from the end

of the lode through the crosscourse to the heaved portion
of the lode, in such a way as to demonstrate that the
dislocation took place after the north and south vein had
been filled with earthy minerals. If it were certain that
none were in existence until after the formation of the
various east and west systems of dislocations, it would
still be true that the different groups of crosscourses
themselves are of many epochs, because they are often
found intersecting each other. These meridional veins
are less undulating than lodes, and seemingly have also
less persistence of character, splitting up into branches,
and dwindling both horizontally and vertically more
than lodes. For this reason though a crosscourse may
fissure an extensive district, its absolute continuity may
well be doubted. There are numerous examples of their
not reaching the surface, and sometimes they are peculiar
to the lode, or even the portion of the lode they intersect.
They are rather less frequent in granite than in slate.

The heaves occasioned by these transverse fissures
harass the miner, and discourages the adventurer, by
cutting off the lode—which is often difficult and expensive
to re-discover,—and by dividing rich courses of ore, the
part disrupted being seldom met with again. Though
perhaps the faulting along the line of cross fissures are
individually of comparatively small extent if separately
considered, yet as they occur with downthrows mostly in
the same direction, the magnitude of the cumulative
result is sufficiently important. Taking for example, the
series of faultings—which have been so well observed by
Mr. J. H. Collins, F. G. S.—along the littoral from the
Saint Agnes Beacon to Padstow, eastern downthrows
are frequently met, some of which disclose slips of more

than five hundred feet, so that the total amount of the dislocation must exceed 10,000 feet. The downthrow does not often take place along a simple fracture, but numerous lateral veins, enclosing slender plates of rock, continue the displacement in the same direction. Owing probably to the thickness of the superincumbent strata and compactness of the rock, the short fragments of the lodes are often surprisingly perfect. The general eastern dip and downthrow of the faults seem to indicate that the Beacon Hill has attained some of its relative altitude by being forced up along their walls. It is then evident that the slates composing the Beacon Hill must be the oldest of the series disrupted, an opinion that receives some confirmation by the proximity of the subjacent granite, which in Cornwall often carries up the slates bodily.

There is a remarkable tendency in the crosscourses, to assume a rectangular position with regard to the group of lodes they intersect. Their strike oscillates between N.W. and N. by E., which is a much smaller range of direction than is occupied by the east and west veins. The variation of crosscourses and flucans from the perpendicular is much less than that of lodes, and their irregularities are not of so pronounced a character. Their dip, equally east or west, is rarely if ever reversed, and the average hade is about 80°. Change from one rock to another does not seem to be attended with such deep inflections as those to which lodes have been subjected— though this may be due to their crossing the strike of the strata—nor, from their lack of metalliferous ores, does it produce any notable change in their contents. As the cross veins are more vertical, and their flexures assimilate more to the mean planes of the fissures than those of lodes, their size is somewhat more constant. The width varies from one foot to five fathoms, and is perhaps a little greater in granite than in slate. The mean width of the crosscourses of the county—quoting Mr. W. J. Henwood—is four feet.

The **structure** of crosscourses is much more open and jointy than that of lodes, the principal joints usually taking the direction of the divisional planes. There is consequently a very " comby " appearance in the veins of quartz, that go to make up a crosscourse, and their walls are often accompanied by numerous lateral joints or veins filled with quartz, which constantly slipping downwards, developes what miners call "disordered ground." There are two other series of joints that are displayed in the gangue where it is compact, or when it approaches to the character of the contiguous " country;" in the latter case these joints often take the direction of those in the country, but when the filling matter is of a different nature, they are confined to the vein.

The characteristic absence of metallic ores in cross courses, has been generally considered by geologists to be owing to their formation during the comparative tran-quility that followed the intense thermal action which bestowed on the veins due to upheaval their metallic richness. But if this view be accepted, it will render the barreness of these east and west veins associated with meridional veins rich in lead rather anomalous. If it be conceded that vast periods of time, according to finite computation, elapsed between the creation of the diverse groups of fissures, intense vulcanic action might be going on in one place whilst dormant in another; and in this way cross veins associated with any particular group may have been filled with a siliceous gaugue, whilst at some distance a series of east and west veins may have been filled with metallic minerals. The alternations in the lodes themselves from earthy to metallic minerals, render it evident enough that there were periods of quiesence when ores of the useful metals were not in solution, or at least when the conditions admitted of no precipitation.

There appears to be no appreciable difference in the mode by which the gangues of lodes and crosscourses have been accumulated, either chemically or mechanically ; and the filling matter bears much the same relation to the encasing rock, being clayey or quartzose in slate, and granitic in eruptive rocks. It also changes with the character of the latter, being schörlaceous or felspathic according to the predominancy of these ingredients in the granite traversed. They often contain fragments broken off the walls, and acquire a slaty texture where the latter are so large as to extend some distance along the fracture. That crosscourses have been exposed to the same depositing conditions as existed in lodes, is exemplified in many districts, and when metallic ores occur, they are exclosed or dispersed through a matrix of earthy minerals just in the same manner as in lodes. There are few metallic ores occuring in lodes that are not represented in the north and south veins, though tin ore is rare, and where observed found only in small quantities. Cupreous and ferruginous sulphides have, however, been worked in the crosscourses at Ting Tang and Consolidated mines, and in the flucans at Herland. Iron oxides are common in Saint Just, and from the crosscourses in

the granite and slate of the Hensbarrow district, most of
the iron ore exported to Wales was raised.   Zinc blende,
silver, cobalt, antimony, and others, have also been met
with.   When at a distance from granite rocks, they
have yielded abundance of galena, and some of the
richest mines in the county have been worked on these
cross lodes.   The production does not seem however to
be persistent in depth, and lead mines have, compared
with those of copper or tin, a very ephemeral though
prosperous existence.

It has been generally considered, that the facility of
aqueous circulation afforded by the porosity of the cross
veins, has had a most fertilising influence on the lodes
which they intersect.   From their non-metallic character,
they are supposed by many to be of later date, but in
this case, it is rather difficult to understand in what way
they could have had any influence on lodes existing at
the time of their formation.   It is more consonant with
ones perceptions of the mechanical effects of the
movements, which have so repeatedly changed the
relative positions of such a hetrogenous mass of
dislocated strata; to believe that the influence of cross
veins has, by severing the rigid longitudinal strips of
rock, permitted descent of the "hanging walls" of lodes,
and thus enlarged or given existence to cavations, the
opening of which was before, more or less resisted by the
inequalities of the flexures along their walls.   Deposits
of rich ore previously existing have often been divided
by a crosscourse, and the manner in which rich bunches
of ore are found in parallel lodes between two
crosscourses, scarcely supports the assumption that
deposits owe their richness to the circulation set up by

their advent. It may be doubted whether at great depths cross veins have much influence on the rich parts of lodes, but near the surface many deposits that may —for want of a better term—be called accidental, owe perhaps their segregation to the freer circulation in the lodes, consequent on the opening of the meridional veins.

Slides. These fissures are without doubt the latest formed, as they intersect the systems already noticed, and only contain the flucany—that is the clayey matter produced by their motion. Nevertheless, though of such recent period, they were certainly formed long previous to the present configuration of the surface, because they cause no more break in the general contour than do the lodes. The number of slides known is not great, and they can only be traced on the cliffs or seen in the mines. They have not been noticed in granite, or metamorphosed strata, but are confined to schistose clayslates, and appear to fall away from the ranges of granite at a flat angle. They have an east and west bearing, and though usually only a few inches wide, attain at Herodsfoot, a size of three feet ; but even when of this size, their contents are derived solely from the abrasion of their walls. It may be remarked, that almost without exception flucany veins —be they lodes, cross courses, or slides—are indicative of motion along their planes, because the flucan is the result of abrasion. The dislocation caused may be very great, so much indeed, that it has often been impossible to find the heaved part of a lode.

Intersections. To the miner the intersections of the lodes with each other is a subject of the greatest interest and importance ; for often on the result of an anticipated junction, depends the working or stopping of a mine.

The conditions under which veins meet are so complicated, that it is scarcely possible to foretell whether the junction will be favourable or the reverse. In many instances the apparent junction of two or three veins may be due to contemporaneous fissuring, and when the rock is not too much shattered, a good deposit often collects there, should the lodes be of nearly similar dip. Frequently the inter-section of lodes is simple, no dislocation taking place, and no doubt numbers of these have been met with in the mines, but have received no notice. Simple junctions are most agreeable to the miner, because the size of the lode is usually augmented. When two lodes intersect in depth at an acute angle, a course of ore may be awaited, and the smaller the angle included the better; should the angle be large, or the rock be much disordered, ore in large quantity must not be expected.

If such lodes were previously well defined, with one or both productive, a rich deposit may be relied on, and even when lodes, previously unproductive, possess a simi-larity of dip and gangue, the junction generally yields ore. Sometime lodes not merely meet, but coalesce with-out intersecting, and continue together both in strike and dip, and instances have been observed where two lodes of opposite dip run down together. Such junctions, are with rare exceptions, eminently productive, and by doub-ling the width of the lode lessen the expense of extraction. Occasionally three lodes coalesce as at South Roskear mine, where the main lode, the Caunter, and Roberts lode, formed a junction, and continuing together for 300 fathoms, produced the immense cupreous wealth for which Old Roskear was famous. (*Vide* PLATE II).

The intersection and junction of the east and west

DIAGRAM SHEWING HEAVES DUE TO INTERSECTIONS OF
VEINS OF DIFFERENT RELATIVE AGE

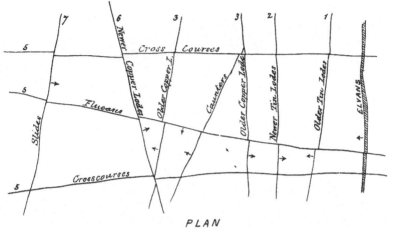

PLAN

The figures shew sequence of vein systems

SECTION

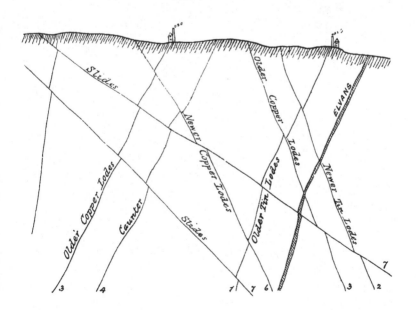

veins, whether in strike or dip, display phenomena similar in degree though not in direction, to those of the lodes and cross veins. In some localities the intersection of lodes of opposite dips has given rise to displacements which are called "upthrows." In Saint Agnes where these dislocations have been developed to a remarkable extent, the lodes dipping south have thrown up the older tin veins, to distances varying from a foot to over twenty fathoms. Wheal Pink lode has in a short horizontal distance been thrice thrown up, so that the same level has passed through it three times, and it appears on the surface as three distinct lodes. These dislocations, as well as the slip of the hanging walls of lodes, seemingly increase in amount with remoteness from granite. Mr. W. J. Henwood mentions that lodes have been heaved by each other as much as forty feet, and mostly to the smaller angle.

There are some examples of the cross veins being intersected and heaved by the later east and west copper lodes ; they are rare, but where occurring give rise to heaves which are sometimes so considerable as five fathoms. The throw of an elvan-course by a lode is of extreme rarity, but an instance occurs in the cliffs between Saint Agnes and Perran.

Heaves. The faulting caused by the crosscourses are called by Cornish miners "heaves," because they imagined that the separation of the ends of a lode was due to horizontal movement. It may be truthfully said, that notwithstanding the attempts of Mr. W. J. Henwood and others by tabulated observations to attain some general laws that would aid the miner in his search after the heaved end of a lode, the data arrived at are but

G

meagre, and the results of the intersections of lodes with
crosscourses not very intelligible. All heaves are,
without doubt, due to subsidence of one side of a vein,
and the inexplicable discordance attending Cornish veins,
may be attributed to their number and to the numerous
epochs of disturbance. Though near the granitic domes,
some amount of lateral movement down their slopes
probably occurred, the extent of nearly all heaves is due
to the same vertical movements that originated the
faults in the coal fields, which, owing to the distinctive
character of the alternations of sandstone, sholes, and
coal, can be accurately measured. The similarity of the
slates of Cornwall does not give the same facility for
realising the extent of faulting, though on the bold cliffs
facing the Bristol Channel, a vast amount of vertical
movement can be verified, and it may be inferred that
the same discordance extends inland. The granite has
been also displaced by the heaves seventy or eighty
fathoms, as by the County and the Great Crinnis crosscourses, both of which are right hand heaves. Elvan
dykes are unaccountably obtuse to heaves by cross veins,
though there are many examples ; perhaps their size,
compactness and rigidity may oppose an amount of
obstruction to movement, that the more yielding slates do
not possess. Some heaves may not be due to motion, but
either to contemporaneous shattering—which can be well
studied in the vicinity of Saint Agnes—or to the initial
blow developing along the line of least resistance. The
plastic nature of some slates is well illustrated by the
small fractures—many probably of the same age as the
lodes—which by local movements have caused heaves at
one level, contrary to that seen at a lower level on the

same lode. Even in crosscourses of importance, local subsidences of yielding strata must often superinduce contradictory heaves; and these movements may still be going on slowly. That such local disturbances have been the cause of much complication, will readily be admitted by those who have observed the dislocations which have taken place along joints, even when they are so fine that their existence is only suspected by the diverse character of the rocks in juxtaposition.

The faulting of crosscourses is less near the granite, and generally increases in amount in proportion to the thickness of the sedimentary strata resting on it. Heaves are not always as simple as they appear on plans, but are often complex, and associated with subordinate fractures connected with the lode intersected. The end of the heaved lode, as Mr. W. J. Henwood admits, having little in common either in size, or contents, must owe their dissimilarity, as near the granite, to frequent re-opening and filling since the formation of the cross vein, or to the extent of the downthrow. This difficulty of recognizing ends of heaved lodes is further increased by the flucany matter along the walls, and the varying obliquity of the lode and the intersecting vein, both in strike and dip. How great this difficulty is, may be conceived when Mr. Henwood states, that he sees no difference that the dip of a lode makes on the heave. Although the extent of the faulting in the clayslates was mostly greater than Mr. Henwood conceived, yet, as the great bulk of his observations was collected near the junction of the clayslate with the eruptive rocks, where the dislocations were at a minimum, his conclusions with respect to length of heave, and its intimate connection with the size of the veins, may still be quoted.

The movement taken by the faulted portions of the strata is similar in direction to those of the coal fields, that is the hanging wall slips down the plane of the fissure towards the greater angle. From a miner's point of view, most of the heaves are to the "right hand" along the crosscourse, those to the "left hand" being to a less distance; and though—following Mr. W. J. Henwood—there are three per cent of exceptions, the general rule is, that the same crosscourse heaves all the lodes in the same direction. The preponderance of heaves to the greater angle is stated by the same authority to be as five to one, and the mean distance of dislocation to be fifteen feet. The proportion heaved towards the greater angle bears a relation to the angle of intersection; for while they are variable when the angle is acute, they are almost all heaved to the greater angle when the veins are nearly perpendicular to each other. Neither the kind of rock nor the metalliferous contents of lodes affect the amount of heaves, as they are due to mechanical motion, but the extent of the heave is as might be expected, considerably influenced by the size of the crosscourse or lode, as it is in consequence of repeated slips of the hanging wall, that veins acquire their magnitude. Accordingly, the researches of Mr. Henwoed have conclusively shown, that the larger the crosscourse the greater is the dislocation of lodes, and that if the lodes are also large the heaves attain their maximum distance. Thus the mean heave when the vein and lodes are small, is only four feet, whilst the intersection of large crosscourses by large veins, raises the average heave to about thirty feet. But the angle of incidence also modifies the length of the heave, thus both a small angle and a right

angle are less favourable than an open angle, the greatest
heaves occuring when the angles of intersection lie
between 30° and 60°, the maximum average heave being
found at an angle of 45°. Yet though the most extensive
heaves are at this angle, they are most numerous at
right angle, and when the angle is very acute heaves are
unfrequent. The amount of heave varies much at different levels, owing perhaps to local movements and to
the flexure of the lodes crossed, whilst the amount of
dislocation caused by a crosscourse is increased, when
by containing clayey matter it becomes a flucan. Even
such a large compact lode as Dolcoath is slightly affected
by veins of flucany matter.

Elvans, though resisting successfully the movements
caused by crosscourses, have been frequently slightly
heaved, and at some intersections the heave has been
important ; for instance, at Trevellas Porth, Gover elvan,
and near Marazion. Elvans have in rare instances
appeared to heave lodes, as at Polgooth and at Boskilling;
it may well be, however, that the initial blow was not
strong enough to fracture the elvan.

It would then appear, that the productions of any vein
systems depend on reiterated dislocation, and these
having been most prevalent near the junction of the
granite with the slate, the best deposits have naturally
become segregated there. The above epitome of intersections will not be very intelligible to the general
reader, nor could any description well be, for it possesses
even for the miner much of obscurity.

CHAPTER III.

Distribution and Paragenesis of Minerals.

The earthy minerals yielded by the veins of Cornwall
have much in common, as quartz in various forms occur
in all the systems of fissures. The non-metalliferous
portions of crosscourses and lodes are generally made up
of quartz more or less crystalline, which in granite
becomes mixed with felspar, schörl, and mica. In some
places where the slates have been highly metamorphosed,
beautiful and rare crystals of siliceous minerals—axinite,
garnet, &c.—are found, but nowhere in such profusion as
in the siliceous greenstone rocks which skirt the coast in
the Parish of Saint Just. Topaz and beryl have been
seen at St. Michael's Mount, and numerous scarce earthy
minerals have occasionally been found in the neighbourhood of Penzance, such as actinolite, thallite, &c.

Cassiterite, or tin ore is scarcely ever totally absent where
metamorphic rocks prevail, and it is consequently seldom
discovered far distant from the granite. Tin ore has been
extensively mined in Saint Just, Saint Ives, and Lelant.
The mines around Tregoning Hill have been very productive, as also those in Wendron, but the deepest, richest
and most extensive deposits of cassiterite, exist along the
northern slope of the granitic hills of Carn Brea and
Carnmarth. An important belt of tin producing strata,
extends through the parish of Saint Agnes to the south

of the granitic ridge lying between the Beacon Hill and Cligga Head. Much tin ore is found in the lodes scattered about the western moiety of the Hensbarrow granite, and there is also a tin district skirting the south flank of the Bodmin range, between Caradon and Warleggan. The cone like granite hills of Hingston Downs and Kit Hill are very productive of tin ore. It is somewhat peculiar that tin oxide is associated with minerals that, like schörl contain a notable per centage of fluoric acid, which is so constantly present, as to suggest that the deposition of stannic ores have been thereby influenced. Although rich mines of tin ore have been known in slate, when elvans are in proximity, it is a mineral which affects crystalline rocks, and has probably been deposited during abnormal temperature and under high pressure. It is always found in east and west lodes, and is not found in cross veins, indeed it is only in rare cases that veins crossing a productive lode rectangularly contain any metallic minerals, but such phenomena have occurred, as at Redmoor near Callington.

Chalcopyrite, or yellow copper ore, is more widely distributed even than cassiterite, but although equally dependant on metamorphic action, it has been deposited in the massive slates which repose on the granite. Like the granite loving tin, copper ores are occasionally deposited outside the strata it most affects, but though rich bunches have been, as at Tresavean and Penstruthal, worked in the granite, yet a copper lode entering that rock generally changes to tin, nor do cupreous ores ever reach the depths attained by stannic oxide. The sub-sulphurretted ores of copper are found in connection with

altered rocks, and prevail in the greenstone near the
junction, notably in Carn Brea and Gwennap districts,
and in the greenstone skirting the western coast of the
Land's End district, whilst the yellow ores occur
mostly in slates more distant. It is therefore not
surprising, that the mines of Camborne and Illogan
which formerly produced copper should now be worked
for tin. The mines of Cornwall on this account yield at
present no great aggregate of copper ores, but formerly it
was raised from nearly all the mines, and most largely in
Gwennap, Camborne, Redruth, Gwinear, Caradon,
Tavistock, and Saint Austell. Wood tin has been
obtained from Penberth, Garth in Illogan, and near
Saint Blazey, but only as specimens.

Cornwall formerly produced large quantities of
Galena, but latterly the mines have become exhausted,
and the low price induced by the vast quantities of bullion
obtained from the Nevada mines have so depressed the
market, that there is but little encouragement to develope
new mines. Lead ores were formerly supposed to be
confined to veins having a meridional direction, but the
discovery of the Chiverton lodes, which had an east and
west bearing, corrected this assumption. The associations
of plumbiferous veins with slates, still further removed
than those of copper from granite, is unquestionable ; but
that a considerable amount of metaphorism is necessary
to insure productiveness, is shown by the meagre and
barren lead veins in slates that are comparatively
unaltered. Small quantities of lead ores have been
obtained from Saint Erth, Phillack, and Gwinear, but
lead lodes are of no value west of the Carn Menellis
range. The principal deposits of lead were discovered at

Wheal Golden and Penhale in the Parish of Perranzabuloe, at Wheal Rose on the coast south of Helston, in
the East Wheal Rose district near Truro, at Trelawney
and Mary Ann mines in the parish of Menheniot, at
Herodsfoot in Duloe, and more lately in the mines
worked on the celebrated Chiverton lode. Galena has
also been raised in many parishes in the vicinity of
Padstow. The Cornish lead ores are much esteemed for
the large proportion of silver they enclose. The most
argentiferous galena has been raised from the Chiverton,
Trelawney, and Herodsfoot mines, and also from some
lodes in the north of Cornwall. Very rich silver lead
ores have been obtained in large quantities from the
Tamar lead mines in the parish of Beerferris.

Silver.—Beside the silver diffused through the
sulphuret of lead, large quantities of the pure metal have
been found in arborescent and reticulated forms. Much
native silver was obtained from the Druid's lode in
Dolcoath, splendid and large specimens from Wheal
Ludcott near Liskeard, at Wheal Brothers near Calstock,
and in other localities. Silver has been found in nearly
every mining district in Cornwall, though generally in
small quantities. Outside of these sources of silver, it is
widely distributed in the gossans and lodestuff of many
groups of lodes, and the importance of these ores—which
are often thrown over the burrows—is very insufficiently
recognised. The present generation of miners read with
surprise, that our foretinners threw away as worthless
many rich ores of copper, but it is possible that posterity
will be still more astonished at our neglect of the argentiferous wealth concealed in the backs of the lodes, and
pitched over the " tip " as " deads." In the Callington

and Calstock districts especially, immense burrows
containing several ounces of silver to the ton which have
been extracted from almost inexhaustible lodes, lie
scattered over the downs. At New Great Consols near
Callington these argentiferous ores were successfully
treated some years since.

Iron.—The ores of this metal, either in the
condition of oxide or sulphide, are always associated with
Cornish ores, and probably no vein is destitute of
ferruginous substances. Its sulphuret is in frequent
association with the principal metallic ores produced, but
it possess no value in this state; when however the ferric
sulphide is unmixed with other ores—as is generally the
case in the crosscourses—they have acquired a value by
oxidation, the sulphur having been carried away by the
carbonic acid contained in the pluvial waters, and
replaced by the proper oxygenic equivalent to form
Hematite, or when hydrated, **Limonite.** Of course
these iron ores can only be formed near the surface where
atmospheric influence can have free action, but the
accidental character of the country where they prevail
has greatly tended to develope oxidation, and to afford
facilities for the cheap extraction of the mineral. Iron
lodes have been thus produced in great number and
importance in the Hensbarrow granite, more especially
in the contiguous clay slates. Many mines have been
opened in the parishes of Saint Enoder, and Saint
Stephens, and along the flank of the granite range to
Lostwithiel, where the vast excavations made on the
Restormel iron lode are indicative of a large and
profitable production. This lode has been thrown down to
the east several hundred feet, hence the size of the vein.

Hematite has been also quarried in the parishes of Saint Wenn and Boscastle, whilst Limonite has been found at Constantine and Lanivet. Carbonate of iron, or Spathic ore, exists in immense masses in the Perran iron lode at a depth of about fifty fathoms, but the backs have been oxidised into hematite and limonite and are worked extensively. **Chrome Iron** has never been found except in very small amount in the serpentine of the Lizard. Phosphate of Iron has been seen in the mines of Wheal Owles, in Saint Agnes, and in some of the mines of the Wheal Jane district near Truro. **Vivianite,** in beautiful indigo crystals two inches long, was found on magnetic pyrites, and earthy phosphate was often met with in some of the Saint Just lodes. **Graphite** has scarcely ever been seen, but is said to occur in an elvan at Restronguet Creek. **Wolfram,** though only existing in masses in a few places, is somewhat generally distributed, and where found is always associated with tin ore, from which, on account of its nearly equal specific gravity, its separation is almost impracticable. It occurs in profusion in the granitic rocks of Cligga Head, and contaminates the ores of East Pool and Drake Walls mine. At the latter place the ores were difficultly saleable until Dr. Oxland installed a process by which tungstate of iron converted into a marketable product as tungstate of soda. Wolfram is also found in many of the tin mines of Hensbarrow, and less conspicuously in some of the mines to the north and east of Carnmarth Hill, and in many others situate at the northern part of Carn Brea. The titanate of iron called Ilmenite is found as a black sand in Manaccan, and massive at Porthalla, and also at Buckland monachorum in Devonshire.

Many rarer minerals are sparsely scattered through the lodes of the county, amongst which may be mentioned an ore of Uranium called pitchblende found in St. Ives, Saint Just, Illogan, and Saint Austell. **Bismuth** has been remarked in Saint Ives Consols, Dolcoath, and a few other mines. Molybdenite in Saint Ives and Gwennap. **Barytes** in Consolidated and Wheal Friendship mines Gwennap, in Saint Austell in the lead mine of Mary Ann, and in the Restormel iron mine. **Cobalt** ores have been noticed at Pengreep, Wheal Sparnon, and Polgooth mines. There are many other rare and beautiful minerals occurring with the useful ores, which, being too numerous to mention, the reader is referred to the comprehensive handbook of Mineralogy, compiled by Mr. J. H. Collins, F.G.S.*

Pyrites, Mispickel, and **Zinckblende,** are found associated with each other, and with the ores of copper, tin, and lead, throughout the mining districts; it would therefore serve no useful purpose to give the localities in which they occur. Mispickel and " black. jack " are especially associated with tin lodes, and less persistently with those of copper. Mundic is often present with tin ores, but is seldom absent from copper lodes, and yellow copper ores are always found chemically combined with a large proportion of the ferric sulphide. Zinc blende is scattered indifferently in lodes, whether of tin copper, or lead, and often all three of the above mentioned minerals are found associated in the same deposit.

Productiveness of Strata.—At the risk of some repetition, it will be desirable to epitomise the lead-

---

* This valuable and portable book of reference can be obtained of Lake and Lake of ·Truro.

ing circumstances affecting the association of metallic ores with particular groupings of strata; for, a very superficial acquaintance with the rocks of the County is sufficient to show the intimate connection which exists between them.   The fossils found at Crinnis beach and other places in rock fissured by lodes, afford direct evidence that animal life flourished ages before their formation.   Although the dependance of good initial fissures on the original mechanical structure of the slates, and on the texture imparted by metamorphism is so absolute as to render hazardous the assertion that the vein systems richest in metallic minerals were formed under the greatest thickness of sedimentary rock, yet, it is nevertheless worth remembering, that those beds believed to be relatively the oldest, have furnished the largest supply of metalliferous products.   A reference to the coloring of the geological map which accompanies this article, will render the association clearer to the reader ; there, pre-Silurian strata—probably Cambrian— are seen to spread over the area in which the famous tin mines of Camborne, Illogan, and Saint Agnes have been worked, while the richest copper mines are excavated in the clayslates of the Lower Silurian.   But notwithstanding this coincidence the best veins are dependant to an important extent on the original structure of the slates, and especially to the crystalline and compact character they have acquired by metamorphic action during the epoch of the granitic upheaval.   In consequence of this action, which, owing to the weak conduction of schistose slates, did not penetrate to a very great distance, the rocks productive of the useful ores are not found distant from granite.   If one could discern the underlying

eruptive rock, the revelation would possibly confirm the
assumption of the non-existence of large deposits more
than a few thousand feet from the granite. At least the
old theory of congenation is shown to be untenable by
the absence of ores in clayslate remote from crystalline
rocks. In plombiferous districts remote from them, as
for example that of Chiverton, which is situate mid-way
between the ranges of Carnmarth and Hensbarrow,
granitic dykes come to the surface, and afford positive
proof of the proximity of the granite beneath.

The chief effect of metamorphism on the structure of
the slates, has been to render them crystalline, and to
harden them into thick alternations of beds different in
density. These characters aided by the enormous pressure,
developed conditions highly favourable to the formation
of fissures persistent in length and dip. There is reason
for conceiving, that the strait smooth foot wall of strong
lodes, is due to the fissure taking the path of least
resistance offered by the divisional planes. There are
other conditions which contribute to the making of bene-
ficial vein groups; for instance, the position and
inclination of the strata resting on the granite is all
important. In the richest districts, the slates are
regularly and moderately inclined, their strike parallel to
the eruptive range and approximately so to the lodes
themselves. This is not the place to speak of those strata
which influence most favourably the productive character
of a lode, as there are very considerable local differences,
but comprehensively, it may be remarked that rocks
should not be of such excessive hardness as to cause a
splintery fracture, nor so soft that the walls would be
liable to contraction by pressure. The connection

between substantial strata and productive lodes, is sustained by the occurrence of ores in connection with ranges of greenstone. A dry shelfy rock, thin soft shales, greatly curved or contorted strata, or slates abutting against the granite at a high angle, are unfavourable to the developement of persistent fissures. It has already been noticed, that the opening of fissures in this alternation of different rocks, has been owing to the action of gravitation impelling the hanging wall to slide down at each repetition of subterrene shocks.

The metamorphism of rocks is not peculiar to the slates, for during consolidation, the latter reacting on the granite produced transition rocks which have a structure and composition different from either the crystalline or sedimentary strata. Schörl gives active assistance in this alteration. As in the slate so in the granite, good fissures and large metalliferous deposits, are seldom, if indeed ever, found outside the metamorphic zone; as distance increases, so lodes disappear, until in the interior of the granitic bosses, the slates remote from granite are not more barren. The metamorphic granite rocks favourable to productive deposits are, like those of slate, variable when one district is compared with another, but generally miners seem to prefer a rock to which a certain amount of decomposition has given a distinctive haziness to the outline of the felspar with a "plumb" texture; whilst a very hard fine grained rock with defined porphyritic character is not regarded with favour. The examples of a lode becoming poor or rich, and altering its metalliferous character, on encountering a distinct change of strata, especially when the change is from granite to slate, are numerous and striking, but the multation may possibly be due to different epochs of deposition. Tin lodes and basic cupreous ores, exist in harder and more siliceous rock than copper pyrites or galena. Thus the testimony of the rocks is in accord with experience in regard to the general distribution of ores, viz.: that copper cannot be expected to occur largely in a tin country, and that lead must not be sought for in a rock congenial to copper, and the converse of this holds good. When one reviews the numerous chain of favourable circumstances neccessary for the origin, development, and filling of productive veins, no surprise need be expressed, that the difficulty of discovering rich deposits in them, demands such patient perseverance, and so much expenditure of capital. First the original composition and structure of the slates, and then metamorphosis into the thick lamellar masses of alternating density requisite for the formation of true fissures; the resting of these more or less crystalline beds on the granite at angles favourable to the development of cavations, with a strike approximately coincident with the trend of granite, and the direction of the line of volcanic upheaval. The continuance of shocks to reopen the fissures, in order to allow faulting between the walls, without which thermal circulation would be languid, and no lodes of any size could develope; and finally the formation of crosscourses to tolerate freer movements of the hanging walls of lodes between them.

Denudation.—The sedimentary beds which rested on the granite at the time of its consolidation, were lifted up by the upheaval, and subsequently abraded and washed away by the ocean, or denuded by atmospheric influences. The metamorphic rocks, which doubtless covered the eruptive

Schorl Rock. — ROCHE ROCKS.

Tabular Granite. — CHEESEWRING.

domes, have also, together with any metallic deposits with which they were associated, been denuded by the waves of the ocean and now form beds of shingle and sand in the English and Bristol Channels. Owing to the disappearance of such a thickness of crystalline rock, the central portions of the granitic bosses have been stripped of much of their metallic wealth by the unbottoming of the elvans and veins. There are however some patches of metamorphic rock still remaining, as in the mining districts of Wendron and Roche; they are of a rather sporadic character, and the larger and more elevated tracts of Bodmin and Dartmoor are entirely barren of the useful minerals, and the only evidence of former fertility— the superficial deposits of stream tin—have long since vanished in ministering to the wants of earlier civilizazations. Formerly, when much of the county was protected by primeval forest, the disintegration caused by the exposure of the rocks to alternate rains and sunshine was less than at present, though now the cultivated portions resist with much pertinacity the degrading influence of the weather. Torrential rains are rare in Cornwall, as the rainfall is distributed over about seven months of the year. The results of the decomposition initiated by aqueous causes is very apparent in the hilly lands, where large blocks of granite have weathered into tabular forms. Many of these old cairns and tors have assumed the most grotesque outlines, for the fashioning of which the Druids have been partly accredited. The high ridge of Carn Brea is one of the localties where the mode in which granite degradates may be best observed. The exfoliation of rocks commences along the jointy structure, down which permeating waters attack the

H

felspathic constituent, and gradually enlarge the veinletts *pari passu* with the denudation of the surface. By this means the most resistant rocks became gradually prominent and their angles rounded off, until in the course of ages the blocks acquired a flattened spheroidal exterior, and were piled one over the other like cheeses.

Lanyon Cromlech.

In some districts the granite is found decomposed to great depths and over extensive areas, and some large tracts of the surface are covered with growan. Where schörl is present the softening of the felspar is very complete.

In the clay slates, the same yielding of the softer rocks to denuding action is apparent where vallies have been eroded around harder strata. Miners have observed that when a ravine or depression runs across lodes, they are often found productive on one side only; and this sterility seems in most cases to be due to change in the structure or induration of the rock, the river scouring its passage through the rock most easily abraded.   The annual rainfall in Cornwall varies from about 35 inches in the west to 60 inches in the north.

Raised Beaches.—The uplifting of the rocks of Cornwall was so slow, as to allow of enormous erosion by the beating of the waves against the coast.   There is no evidence of any paroxysmal event, and nature appears to have enjoyed a tranquility that permitted an exuberance of animal life and vegetation, equal to that now existing. Though it is seldom that a casual observer would be able to find positive proof of the gradual rising of the granite from the ocean, owing to the many causes which combine to degrade rising cliffs along the coast line ; yet there are numerous places, where, owing to favoured situations, fragments of sea beaches raised from 10 to 60 feet above sea level have so far resisted obliteration.   These beaches—geologically recent, but of great antiquity— prove by their general appearance and contents, that the land continued to gain elevation down to a period when the species of marine life had assumed their present forms. The action of the weather, and the accumulation of detritus, causes a talus that sometimes reaches the sum- mit, which has concealed many of the ancient beaches, so that in some thousands of years they may be unrecogni- sable.   The raised beaches are composed of layers of boulders, pebbles, gravel, and sand, analagous to those which now so beautifully adorn the crescent shaped margins of the bays that engirdle the coast.   They are also bestrewed with white limestone and basalt, and with flints similar to those found on the downs of the Land's End.   They extend down nearly to

the present beaches along the shores of Mousehole and
Saint Just, where there is an immense thickness of granite
boulders.   Bordering the south western flank of the
Bodmin granite near Saint Neot, and on towards
Warleggan, there are somewhat similar accumulations of
granite boulders, amongst the lower stratum of which
cassiterite has been extracted.   Where the miners have
not made excavations, this detritus is so completely
concealed all along by a talus of debris, that the extent
of the remains of this ancient beach cannot be ascertained.
It is not often that the granitic islands would be able to
show such evidence of the sojourn of the Atlantic waves
around their bases, because the denudation of perhaps a
thousand centuries could scarcely be expected to leave,
except by a succession of very favourable circumstances,
any vestiges of sea shore.   As these raised beaches are
found numerously along the coast, both in the granite
and clay slate formations, it is needless to enumerate
them.

Submarine Forests.—The long continued elevation
of Cornwall, which the boulder and gravel deposits inland
at Saint Neot, and along the coasts, show to have
happened, has in the newest recent epoch, given place to
a movement of subsidence.   Submarine deposits of tree
stumps, vegetal remains, bones, stanniferous gravels, &c.,
have been found at the mouths of nearly all the creeks,
both on the north and south coast, at a depth which has
been shown to reach thirty feet at Pentuan and other
places.   The filling up of the creeks with mud, and the
general shallowing up of the English and Bristol channels,
also point to the cessation of upheaval movements.
Within the past half century, the debris sent down the

streams from the tin mines and china clay works, have
filled up many important creeks, such as Tregony, Gweek,
Devoran and others, and the lapse of another century
may find many Cornish estuaries changed to verdant
pasturage.

CHAPTER IV.

THE MINING DISTRICTS.

The general dispositions of the vein systems of Cornwall, and the occurrence in them of useful metalli-ferous products having been noticed, a clearer conception of the *tout ensemble* may be acquired by the reader if a few sentences be devoted to describe the prominent points of the groups of mines lying around each granite boss.

THE SCILLY ISLES.

Though the character of the rock between Scilly and the Land's End is concealed by the ocean, one can without hesitation believe that like the eruptive bosses of the main land, the numerous isles and rocklets are simply the tors of a submerged tract of granite which has a quaqua-versal dip beneath clay slates similar to those reposing in the other granitic ranges. Although the rocks of Scilly appear to possess the cubical and columnar structure of the Land's End district, it is, as far as known, unproductive of metallic minerals. Possibly previous to denudation, clay slates may have covered the plutonic rock, and a zone productive of minerals may have been existent ; if this were so, the scarce patches of elvans, and the occurrence of some small veins, in which were occasionally discovered traces of

tin, copper, and lead, may represent the rocks which at
the time of the vein formations were in a state of
viscidity that rendered the retention of the useful ores
improbable. The few veins that can now be seen, have
no importance and produce no metallic minerals.

ST. JUST DISTRICT.

There is no group of mines in the whole of Cornwall
and Devon which presents such varied interest as that of
St. Just, whether one considers the opportunities afforded
for rock study by the deep zawns which so frequently
penetrate to the junction of the granite with the sedimentary strata, the remarkable systems of fractures, the
development of curious and rare metalloid minerals, or
the distance to which the miner has burrowed beneath
the sea under the shadow of the wild, rugged headlands,
amongst which the houses and engines of many of the
mines are so dangerously and singularly perched. The sea
having worn its way through nearly to the eruptive rock,
has left but a fringe of highly siliceous slates resting on
it, and along this narrow band the copper and tin mines
have been worked. The true crosscourses, which are but
few, have a bearing approaching that of lodes in other
localities.

Fine crystals of orthoclase and white mica near the
junction, and large patches of black tourmaline are
frequent. Pinite is also prevalent at Nangisel Cove in
Sennen, Tol-pedn-Penwith near the Logan Rock, and at
Lamorna and Mulvra Hill. Axinite has been found at
various spots around the Land's End, but is best displayed

in the bold rocky cliffs of St. Just, amid which at Huel
Cock a vein of violet crystals was formerly worked for
specimens.   At "The Bunny" are some peculiar floors of
cockle or schörl.   The metamorphic rocks which lie on
the granite along the St. Just and Zennor shores are very
complicated, and will well repay study,   They consist of
slaty micaceous felspar rock, usually found next the
granite, granular felspar rock, a hard siliceous green rock,
and hornblende rock.   These rocks are generally very
characteristic, and sometimes reciprocate with each other;
they are well developed at many places, but probably
they can be best observed at the Gurnards Head.   The
altered slates, which are very hard and weather well, incline northwards from 25° to 45,° rarely the dip of the
junction is reversed over small spaces where the metamorphic strata seem to support the granite.

The occurrence of clayslates along this coast, or even
anywhere near the margin of the Land's End boss is in-
frequent.   The St. Just veins pursue very abnormal
bearings ; the principal lodes approach north-east, a
direction nearly parallel to the fissure systems of Germoe,
others are more nearly east and west.   A peculiar group
of lodes, locally called guides, which have some characters common to cross courses and the same bearing as the
cross veins of the Caradon district, often curve away
westward and run along with the lode, and at these junctions rich courses of ore have been successfully worked.
The lodes cross the border of the granite at right angles,
and appear to lose their productiveness for tin towards
the interior, consequently the junction rocks which are
the most productive, have been followed several hundred
fathoms westward under the sea.

The lodes have not been so rich as those of some other parishes, but they have been fairly productive. Although Levant returns 1000 tons of copper ore yearly, most of the mines are worked for tin ore, which being found in the granite near the junctions, the deep workings are moving slowly westward. The richest tin mines are in the Botallack group, and here the veins cross and coalesce with each other in a way quite unique. The richest copper ores found in the county were raised at Levant, Botallack, and other mines, in the metamorphic rocks along the junction, but the lodes changed to tin on penetrating the granite. The ores were vitreous, native, with copper pyrites, and some salts of copper, the ores sold averaged over twelve produce.

Balleswidden is a mine in the granite, which has been worked on a course resembling an elvan more than a lode. The matrix is composed of felspar, quartz, white mica, and schörl, amongst which cassiterite is distributed irregularly in large quantities. In the junction slates, the gangue of the lodes is quartzose with some ferruginous matter, and occasionally some felspar clay and lime ; in the eruptive rock the vein stuff is also quartzose, but a granitic character predominates. Schörl, chlorite, and brown iron, are met with in most of the lodes, and sometimes fluor, mundic, and ologiste iron. In the guides and scorrans, quartz and ferruginous oxides prevail. Mispickle or blende is but little associated with the ores in the matrices of the St. Just lodes. Several of the rarer ores of iron are found in small quantities at Levant and Botallack in the form of arseniates and phosphates. Native bismuth and its oxide and sulphide are found in the same mines, and at Balleswidden, together with apatite, fahlerz, and ores of cobalt, nickel and uranium. Some carbonate of bismuth raised at Wheal Owles was sold at the rate of £130 per ton. The lodes in Morvah and Zennor were in similar rocks and produced much the same minerals. The cobalt ores of Wheal Owles were sold for £40, whilst small quantities of the ores of uranium (pitch blende) fetched over £100 per ton. The situation of the richest and most extensive mines being on the brink of the lofty accidented cliffs, much of the machinery, engines, whims, and roads, are built on over-hanging rocks, or on the summits of vast cliffy precipices rising from the ocean, and communicating with the shore by frail wooden stagings. At the Botallack mines, the difficulty of transporting the ores from the ends of the deep levels extending a thousand yards under the sea, compelled the adventurers to construct a diagonal shaft seaward in order to afford a more direct communication with the bottom of the mine. Down this incline Queen Victoria descended about the year 1860, and only a short time subsequent to her visit, the rope attached to the wagon snapped, and several miners were hurled down 200 fathoms and killed. At Levant Mine they have a steam engine at work, 1250 feet under the surface, situated 600 yards from the coast, under the Atlantic billows.

ST. IVES DISTRICT.

The geological character of the rocks are nearly the same as in the parishes just noticed, the same bands of felspar and hornblende rocks, skirt the granite to St. Ives town, and thence eastward to the Wheal Providence

mines.   The mines usually produce tin along the littoral
from Boscaswell to St. Ives Consols, but they have, with
few exceptions, been unsuccessful.   The country is wild
and sparsely inhabited, but the uplands are rendered
romantic by numerous cairns of great elevation that rise
from the heath-clad downs, the side of which are bestrewed
with enormous granite boulders.   The transition of the
granite to slate is not abrupt ; the former grows fine-
grained and quartzose with schorly spots, graduating into
a rock composed of schörl and quartz, which is succeeded
by felspathic rocks similar to those of the coast to the
south.   The crystals of felspar are large, and often manifest a remarkable coincidence of direction.

The lodes bear east-by-north, and the trawns—as the
cross courses are called in this district—are nearly perpendicular to them.   The matrix of the ores and the ores
themselves, are much the same as in St. Just, rich copper
ores being found along the junction in the greenstone
rocks and cassiterite in the granite.   The walls of the
trawns are argillaceous and smooth, and their principal
contents are quartz, schörl, and earthy iron.   In the
parish of St. Ives occur the remarkable deposits known
as " carbonas," large masses of stanniferous granite that
assume a sort of linear direction and are obscurely connected by the jointy structure of the rock with lodes.
One of the carbonas in the St. Ives Consols mines was
250 yards long with a diamater of about 20.   It had
only a very partial attachment to any lodes, but appeared
to be a portion of granite into which the tin ore had been
gradually accumulated, filling up the joints and veins,
and substituting itself for some of the constituent parts
of the rock, which consisted of the usual granitic minerals,

felspar, quartz, and schörl. The deposit shaded off into the granite, becoming gradually less and less stanniferous. Enormous quantities of tin were raised from these carbonas, which are not developed in any other part of the county. They do not appear at the surface, but some very similar deposits in Towednack and Zennor have been worked at " grass." The copper ore raised from Huel Trenwith and Providence mines, was contaminated by the oxides and phosphate of uranium. Specimens of bismuth have been picked from the ores of St. Ives Consols, and cobalt bloom has been seen in Huel Trenwith. Although the population of St. Ives was two decades ago actively engaged in the flourishing mines, there is now not a single mine at work, excepting a few tinners who are employed in St. Ives Consols above the adit.

LELANT MINING DISTRICT.

The lodes of Towednack and Lelant are parallel to those of St. Ives, as are also the cross courses. Outside of those worked in the greenstone of St. Ives, all the mines are tin producing. The Huel Margaret group is situated on the junction of the granite and slate, and some years ago great profits were made, but the mines are now only tin producing, and pay no dividends. The Huel Reeth mines, wholly in granite and remote from the clayslates, were formerly very productive, but failed in depth. The gangues are granitic in the eruptive rock, consisting principally of quartz, chlorite, brown iron ore and sometimes mundic. The trawns are composed of schörlaceous granite, and some of the cross courses are made up of combs of crystalline quartz. The gangues besides tin and copper, enclose sometimes in small quantities blende, mispickle, and calamine ; of the rarer metals, molybdenite has been observed in the ores of Huel Mary. Mining is at a very low ebb in this district, and the three or four mines returning tin are making no profit.

MARAZION DISTRICT.

No mining of any importance has taken place in the interesting alternations of slates and greenstones, which occupy the deep bay in the granite immediately west of Penzance. Several fine elvans traverse these strata, and enter the granite. In the southernmost of these, generally called the Penzance elvan, which crops out of the sand about a hundred fathoms in front of the Esplanade, the jointy structure is filled with tin ore for a width of twenty feet. It was worked by a miner a century ago, who with that boldness and practical skill characteristic of the Cornish miner in all the mining fields of the world, built a shaft in the midst of the sea, which he connected with the shore by means of a wooden staging. The shaft was sunk several fathoms below the bottom of the sea, from which the water was pumped out by flat rods attached to a steam engine on the Green. Nearly £100,000 worth of tin ore was raised, and the adventure was exceedingly prosperous until the staging was demolished by a large ship driving across it during a violent storm. The mine was re-opened at a great expense thirty years ago, but was unsuccessful. Some rare minerals were obtained, amongst them tin white cobalt.

HE most delightful spot in Cornwall, whether for the tourist or artist, the mineralogist or geologist, is the St. Michael's Mount. This conical island, 500 yards in diameter and 200 feet high, is made up half of sedimentary strata and half of granite. It is full of interest to the geologist because of the phenomena associated with the transition rocks, and here the mineral collector can fill his bag with specimens of rare and beautiful minerals, viz. : apatite, beryl, pinite, topaz, garnet, tourmaline, and others, together with bell metal ore, tin ore, and wolfram.

In the parish of Ludgvan, between the towns of Penzance and Marazion, the Huel Darlington group of mines were extensively worked a quarter of a century since. The lodes traversed deep blue slates, sometimes with greenstone and actynolite rock. The principal lodes were highly mineralized, producing both tin ore and copper pyrites in a gangue of quartz, slaty matter, mundic, with occasionally some arseniate of iron. At West Darlington some argentiferous lead, silver glance, and native silver were found.

In the parishes of Marazion, St. Hilary, and Perranuthnoe, the slate in the vicinity of the lodes is grey to dark blue, and is crossed by highly porphyritic dykes which are signalised by the presence of pinite. The gangue of the veins is quartz, slate, felspar, clay, earthy iron ore, and mundic, accompanied by melanterite and copper pyrites in the slate, which is replaced by vitreous copper ore when the lode comes into contact with the elvan course. The influence of the strata on the deposition of copper ore seems well displayed where a lode happens to follow the wall of the elvan, as the vitreous ore is next it whilst the yellow ore is towards the slate. In most of the lodes considerable quantities of tin oxide exist, and formerly black tin was sold in some quantity, though at present, except at Penberthy Crofts and at Tregurtha and Owen Vean, scarcely any mining is being done. Many copper mines were worked in the clay slates of Perranuthnoe; Wheal Speedwell, Wheal Neptune, and Halamanning mine were among the richest.

GODOLPHIN & WHEAL VOR DISTRICT.

The granite boss stretching across the parishes of Godolphin and Breage rises into two eminences, the beautiful cone of Godolphin 532 feet, and Tregoning ridge 635 feet above the sea that washes its southern flank. The junction around the hills is studded with mines, many of them celebrated for their productiveness. The transition rocks are extremely interesting, and very similar to the alternating strata of granite and slate seen in Dolcoath and Cooks Kitchen underground, but which owing to the ravages of the sea, can be seen open to the day in the celebrated Trewavas cliffs. Many of the junction rocks are interesting, because they demonstrate how gradual the change from slate to granite may be; at Carleen, the lamellar dark blue slate, slowly becomes a

crystalline quartzose rock, which, developing into
brownish felspar rock, shades off into the usual kind of
altered granite seen near the sedimentary rocks. The
Tregoning Hill is schorlaceous, and assimilates to that of
Hensbarrow, is prone to decompose and produces china
clay.

The lodes are very numerous on the western side, and
some of them have produced much tin. The celebrated
Great Work mine in the hollow between the two hills,
was worked 150 years ago, and sold considerable
quantities of black tin, though little is being done there
now. The lode, so rich in granite, contained no tin of
value when it entered the clay slates. There is a very
regular group of veins extending from Tregoning Hill
across the slate basin to the Wendron granite, which was
celebrated half a century ago for the extraordinary riches
yielded by the Great Wheal Vor lode. It was re-started
30 years ago, and a hundred inch cylinder and other
pumping engines were erected to drain the mine to the
bottom, which however was not found to be so stanniferous as the reports lead the adventurers to expect. Many
other parallel lodes were rich, particularly Wheal Metal
and Wheal Sozen. It is a most productive locality for
tin, and considering that the mines around Wheal Vor
have paid more than £1,000,000 in dividends, has been
strangely neglected. The rich old copper lode of Godolphin was in slate, and the grey ore cropped out at surface.
It produced various cupreous ores and some stannic oxide,
which were enclosed in a gangue of quartz, brown iron
ore, chlorite, slaty matter, and mundic. The Great Work
mine is in granite, its lodes are granitic with brown iron
ore and mundic, in which were distributed large quanti-

Columnar Granite. — TOL-PEDEN-PENWITH.

Published by E. Besley

ties of tin ore. The rich tin mines of the Great Vor district were in slate, and the gangues consisted chiefly of quartz, slate, chlorite and iron pyrites, which became arsenical at increased depth. Of the scarcer minerals, tungstate of lime is found at Penberthy Crofts, fahlerz at Great Work and Wheal Prosper, mimetite or arseniate of lead at Wheal Prosper, apatite and topaz at Tremearne, whilst pinite is prevalent in the granite of Tregoning hill.

In the clay slate west of the Looe Pool, is a group of meridional lead lodes which produced argentiferous galena in large quantity, half a century since. About the same time the old Trewavas mine in granite was a rich copper mine, and as at the St. Just mine, the engines were erected on frowning granite crags, and the ore was followed under the sea. At present the waves wash into the workings, which are full of sea water and cannot be drained.

The face of the country from Helston south to the serpentine of the Lizard, and along the banks of the Helford creek, is a succession of hill and dale, picturesquely diversified by wood and water. The tabular appearance of the serpentine which forms the Lizard, invests it with rather a monotonous aspect, though many romantic, but short valleys and ravines, have eroded passages through the lofty and majestic coast line, which the Lizard opposes on all sides to the ocean. The numerous deep indentations in the variegated rock, leading to lovely beaches embosomed in rugged cliffs, attract numerous visitors. Very little has been done on the southern slope of Carn Menellis, the only mine of importance which yielded tin being Wheal Vivian. At Gweek some elvan courses cross the creek in a north-east direction, but no lodes have been observed. At Swanpool, near Falmouth, much

I

lead was raised, and an arsenic refinery erected some thirty years since, but the operations were unsuccessful.

Several beautiful and remarkable minerals are found about the serpentine of the Lizard, viz.: steatite, asbestos, diallage, actinolite, and schillerspar. Some copper ores occurred at Polurian cove, Huel Unity, and Huel Downas; magnetite at the Lizard Head and Gwinter; ilmenite at Porthalla; chromite at Cadgwith; and menaccanite at several places in the parish of St. Keverne.

WENDRON MINING DISTRICT.

The rounded eminences of Carn Menellis present a dreary expanse of furze-clad summits, whose partially cultivated slopes shade off into dreary moorlands, whilst the valley bottoms are disfigured by the unsightly stream works of the ancient and modern tinners. The Wendron tin mines are situated mostly in the granite, the mineralized belt extending to Stithians. The principal mines are grouped around Porkellis Moor which is traversed by numerous lodes, but although much tin ore was in the aggregate extracted, the district was not very successful. This was in some measure due to the immense body of water held by the Porkellis Moor compelling the erection of large pumping engines, though it must be admitted that there is a tendency in the lodes to dwindle in size and productiveness in depth. A short distance south of Porkellis Moor, is the Wheal Lovell group of tin mines, long famous for the rich courses—or rather columns—of cassiterite that bestowed a short though dazzling existence to the mines in which they occurred. The deposits of

East Lovell were very peculiar, they were not formed in
the veins, which are usually small, but apparently by the
infiltration of the tin oxide into the walls, which, spread-
ing to a distance of some feet replaced the granite. But
in accordance with the general law, that the hollows produced by movement are the depositaries of metal, these
carbona-like masses have developed where the vein is
largest, and these spaces having but a small angle from
the perpendicular, the aggregations bear a considerable
resemblance to pipe veins. The gangues of the lodes in
Wendron are pronouncedly granitic, and the indications
seem to point to impoverishment increasing with depth.
The only mines returning tin just now are Basset and
Grylls, East Lovell, New Lovell, and New Trumpet Consols, but the aggregate quantity produced by them is less
than a hundred tons.

Among the lead ores raised from the lode of the Huel
Penrose group were mingled some arseniate, sulphates,
carbonate¹ and phosphate of lead. Cerussite was also
found at Huel Unity in Sithney and at Huel Ann in
Wendron, and wolfram in the tin ores of Prospidnick.

GWINEAR AND CROWAN DISTRICT.

Includes the mines between the granite bosses of Land's
End, Godolphin and Carn Menellis. The upland plains
present in a milder form the usual barrenness and dis-
figurement of a mining field, but the low undulating surface, slopes northward through a well cultivated country
to the lowlands, around the muddy creeks of the Hayle
estuary. The elvan courses and greenstone dykes are

very numerous and their direction various. The diverse
bearing of the lodes, and the general dip of the lodes to
the south, tend to make the district remarkable. Through
St. Erth and Phillack the slates are soft in character and
of a deep blue colour, the lodes in which have been productive of copper. Mellanear, which sells about 7000
tons of copper pyrites a year is now the only mine of
importance. The lodes yield beside copper, blende, and
some plombiferous ores in a gangue similar to that filling
the lodes in Ludgvan and Marazion. The rich copper
lode of Huel Alfred produced very fine specimens of
plombic phosphate and other rarer compounds of copper,
lead and silver.

In Gwinear the slates are rather fissile and flaggy, they
have various tints of blue in some parts, at others reddish
brown, and the parish is essentially cupriferous. Though
several important mines existed a quarter of a century or
more ago, there is at present not a single copper mine at
work. Huel Tremayne was extensively worked for tin,
but the only tin mine now at work is Wheal Jennings,
which returned in 1882, only 26½ tons of black tin. The
gangues are more quartzose than those previously noted.
The district is nowhere deeply mined, the shafts seldom
reaching much over a hundred fathoms in depth, the
sinking being stopped whenever unproductive ground
was touched. Tin ore has been raised in small quantities
in most of the mines and was abundant in Trevaskis,
Carzise, and Rosewarne. The historical mines of Herland
and Huel Unity, have yielded a number of rare and
curious minerals in a gangue which is very quartzose and
slaty. These comprise horn silver, argentite, red silver
ore and native silver, bismuth ores, and ores of uranium,

as well as arseniate of cobalt; whilst at Relistian and Trevaskus tennantite, and at Huel Unity and Trevaskus molybdenite, earthy cobalt and bitumen were found. Huel Unity was also remarkable for the variety of cupriferous ores yielded by the lodes. Relistian mine was first worked as a quarry, which is still open for a length of nearly a hundred yards and a depth of seventy feet.

The characteristics of the parish of Crowan are much the same as the adjoining parish of Gwinear, the mines producing chiefly copper pyrites. Formerly those most famous for their metallic wealth were Wheal Strawberry, Huel Treasury, Binner Downs, Wheal Abraham, and Crenver mines; but for many years last past very few mining works have been carried on in the parish. Some years ago the old mines of Crenver and Abraham were drained to work the tin stuff supposed to have been put away to "stull," but no success rewarded the expenditure of the £150,000 required to open them. The mines are all comparatively shallow, and spirited sinking would probably be attended with great success.

CAMBORNE AND ILLOGAN MINES.

The country in these parishes amongst the mines is rendered dreary and wretched by the countless heaps of mine rubbish, the large areas covered by unclean and delapidated dressing sheds, and by numerous mine streams, foul with slime which meander across roads and fields with aimless persistency. The Carnbrea Hill, crowned with magnificent rocks and a tall monument, is the highest land in the vicinity, having an elevation of 740 feet.

At the northern foot of Carnbrea Hill, a few parallel lodes—extending less than two miles in length—have been worked for hundreds of years, and have yielded an aggregate of millions of pounds sterling worth of black tin. Though not so profitable as some few years since, when labour was cheaper and the price of metal higher, the aggregate production of Dolcoath, Cooks Kitchen, Tincroft, and Carnbrea is nearly 4000 tons per annum. The southern foot of this narrow granitic ridge has been also rich, and sells at present more than 2000 tons of tin. Thus a third of the tin produced in the county, is raised from the parallel groups of lodes which follow the direction of Carnbrea ridge. Most of these mines produced copper pyrites and vitreous ores until three decades since, when, after passing through some unproductive ground, the copper was unbottomed, and rich courses of cassiterite were everywhere discovered. A group of lodes a short distance north, worked at East Pool and Wheal Agar bid fair to be very productive, as the former mine sells about 1500 tons of black tin yearly, and is in the dividend list. There is also a group of rich lodes associated with greenstone dykes, which at Roskear and the Setons have sold enormous quantities of rich copper ores. It is impracticable to give here the names of the rich mines which have been worked in these parishes, but speaking broadly, the whole ground is rich in metallic minerals, and wherever a good lode has been perseveringly followed non-success has been rare. The alternations of granite and slate at the junction are most peculiar, the latter having been metamorphosed into thick lamellar dark blue or purple slates, crystalline in character and felspathic in composition. The lodes on entering the granite soon change to tin and usually increase in size, and a very successful future may safely be predicted for many of the mines in the stanniferous zone. The mines are becoming very deep, Dolcoath having reached a distance of 400 fathoms from the surface. The immense gunnies in this mine give cause for anxiety, as in 1828 an immense subsidence took place, and there is at present a slow movement downwards as is evidenced by occasional slips of ground.*

The lodes have an average direction of about 30° south of west, they incline north and south, and being in transition rocks are both productive. The caunters have a bearing about 20° north of west, and have yielded some galena and blende, though chiefly copper pyrites. The gangues of the lodes throughout are much the same, viz.: quartz, brown iron ore, chlorite, and sometimes fluor and lithomarge, which, where greenstone prevails, is mingled with quartzose slate, in the blue silky slates with slaty matter, and in the eruptive rock with granitic matter. In the metalliferous portions of the lode, mundic, blende, and galena are often present. Besides copper pyrites, vitreous copper and purple ore, there are found occasionally native copper, red and black oxides, green and blue carbonates, with specimens of fahlerz, hydrous and anhydrous sulphates and phosphate of copper. The only ore of tin is the binoxide, but small quantities of wood tin were found in the Garth mine and tin pyrites at Carnbrea. Silver or its ores have been found in Huel Basset, East Pool, West Dolcoath, and North Dolcoath. It was also formerly largely raised from Carn Entral mine near Tuck-

---

* In 1687 a superficies of nine acres subsided into the old mine workings at Fahlun in Sweden, and left a chasm 260 feet deep.

ingmill, in which mine it was found native, as sulphide, and combined with antimony. Cobalt and nickel have been raised from East Pool and sold for £40 per ton, they occur also at Dolcoath. Various ores of Uranium —pitchblende, torberite, antunite and johannite—are found in South Basset, Tincroft, Dolcoath, and East Pool; in the latter mine some tons have been sold at prices varying from £70 to as high as £200 per ton. Wolfram occurs with tin ores at East Pool, Carnbrea and Dolcoath; a considerable amount is sold at East Pool at about £12 per ton. Bitumen has been seen in East Crofty, Cook's Kitchen, and Tincroft. Bismuth and bismuth glance have been sold at Dolcoath and East Pool for £50 a ton. Beautiful specimens of the ores produced by the deposits around Carnbrea and Carnmarth, may be studied in the collection at the Jermyn Street Museum, London.

REDRUTH AND GWENNAP DISTRICT.

The mines worked around the circular boss of granite called Carnmarth, although of comparatively shallow depth, have been amongst the richest in the county, second only to those associated with Carn Brea, and it is possible that with pluck and perseverance—similar to that exhibited by Capt. Charles Thomas in 1850, when insisting on sinking the shaft through poor ground during the transition from copper to tin—many of the rich copper lodes might in depth be found to enclose rich courses of cassiterite. But for Captain C. Thomas the mines working in the Dolcoath lodes might to-day be as idle as those of Gwennap. From the summit of the

Carnmarth cone—771 feet above the sea—the mining tracts are seen sloping away to the Truro river eastward, to the coast northward, and to Hayle westward. This magnificent prospect includes the oldest, richest, and most extensive tract of mineral ground in the county, and, on account of the unfruitful character of the soil, and the myriads of burrows of various shades of blue, yellow, and brown, is the most barren in aspect. The surface of the downs is covered with a white layer of quartz fragments, the debris of ages of denudation.

To the south of Redruth Town is a group of copper mines, of which Tresavean and Wheal Buller form the centre, that has been exceedingly productive of copper, both in the slate and in the transition rocks accompanying the junction. Most of these have been exhausted and are now idle; the only mines now working are West Poldice, Tresavean and North Penstruthal. Wheal Gorland has produced much fluor spar, which formerly sold for 22/- a ton, but is now worth only 10/-. To the north of Redruth, the Wheal Peevor cluster of lodes yielded excellent returns of tin for many years, but they have fallen off very considerably lately. The Redruth lodes which bear about 24° N. of East, being a continuation of the Illogan district, are filled with the same gangues, and produce the same common and rare minerals.

The quantity of tin returned from the Camborne and Redruth mines during 1882, the slimes from the dressing floors of which, find their way through the Portreath and Red rivers to the sea, is about 7900 tons of black tin, and a further quantity of about 1520 tons (£65000 worth) is obtained from the slimes dressed by the

streamers on those rivers. More than 75 % of the tin
produced by the county, is obtained within the upper
watershed of these two little valleys. Allowing five per
cent for the tin ore finally lost in the Bristol Channel,
the miners lose from their dressing floors, more than a
fifth part of the tin raised from the mines. This seems
an exceedingly large proportion, but when one learns
that no less than forty-four companies are employed in
incessantly re-washing the slimes along the river beds, the
difficulty of devising any means to check this waste at
the mines will be readily acknowledged.

The **Gwennap Mines,** though some tin was obtained
from them, must be considered as **forming a copper
district,** since all its lodes have yielded that metal in large
quantity. Unfortunately from want of a good system of
mining, and of combination between the numerous
companies, the mines were worked a good deal on the
hand to mouth system, and no provision made for the
future by keeping the shafts well ahead of the stopes;
this system led to mines being abandoned after the
shallower courses of copper were exhausted, and the
water became too fast for the remaining mines to work,
except at a loss, though some of the lodes were rich when
the last suspension of Clifford Amalgamated took place. At
this mine no less than nine pumping engines, averaging
76 inch cylinders, besides nine other engines, were
employed. To re-open the mines would require a large
capital and years of time to reach the deeper courses of
ore, which analogy would lead one to believe must exist.

The County crosscourse traverses the parish of
Gwennap, and is said to be traced from Tywarnhaile on
the north coast into the Carn Menellis granite, a distance

of eight miles.　Some of the lodes running eastwards are said to continue to the Wheal Jane group of mines, which is not less than seven miles.　It may be questioned if the identical fissures are so persistent, though the system of fractures is indubitably the same.　The famous copper mines of Gwennap have been the richest in the county, and the immense dividends declared during the first half of the present century, imparted such confidence to the public, that it has taken a quarter of a century of share jobbing and "calls" to quench the spirit of speculation.　The mines that produced the largest masses of copper ore were worked in the group of lodes running through Cosgarne Manor.　Unfortunately the statistics of sales are not accurately known, but the aggregate value would not be less than £10,000,000, whilst the profits made amounted to more than £2,500,000.　In this district—which employed 5000 persons thirty years ago—the only mine of any importance in operation is West Poldice, which returned in 1882, 143 tons of black tin.

The slates are thick lamellar, of a pale to a deep blue colour, and often silky in texture.　The gangues are composed of quartz, red and brown iron ores, chlorite, and some fluer, which are associated with argillaceous matter in the slates, and with granitic substances when the lodes traverse granite or elvan courses.　The ores produced are chalcocite and copper pyrites, mundic and blende, and near the surface native copper, cuprite, and black copper ore.　The mines on the Poldice group of veins produce much tin, especially in the deeper levels, accompanied sometimes by wolfram; in Huel Unity Wood were collected the finest crystals of cassiterite ever

seen in Cornwall. Of the rarer ores, tin pyrites were
found in the Barrier mine, barytes in the United and
Consolidated mines, molybdenite in Huel Friendship,
cobalt and uranium in Huel Gorland, Huel Jewell, and
Ting Tang, vivianite in Huel Gorland, and bitumen in
Poldice.

The Gwennap elvans and lodes prolong themselves
into the parishes of Kenwyn and Kea, where they produce
tin with large quantities of iron and arsenic pyrites.
Wheal Jane, which is the only considerable mine in
operation, sold 117 tons of black tin in 1882.

SAINT AGNES DISTRICT.

With but a small break, the mineralised slates of
Gwennap extend to the Saint Agnes coast, along which
from the Beacon—621 feet high—to Cligga Head, there
is evidently a subterrene ridge of eruptive rock, that
may perhaps extend itself under the slates of the
Perranzabuloe district. Although the absolute height of
Saint Agnes Beacon is not great, yet as the plateau like
moors, and sub-arid downs which surround it, are
relatively much lower, it is a very conspicous object for
miles around; the cone also possesses much geological
interest on account of a belt of peculiar fossilless sand
and clay which encircles its base. The granite of the
beacon has a character intermediate between that rock
and elvan, and is traversed by numerous veins of tin
oxide. At Huel Coates the large orthoclase crystals
have been often decomposed, and by substitution made
pseudomorphic after cassiterite. The Cligga Head mass

has also many features possessed by elvans, and is made up of many dykes which differ in texture and components, so that in one the rock may be porphyritic and in another resemble greisen. Numerous veins of tin oxide with schörl and wolfram give the whole rock a dark and mottled appearance, and, owing to the burrowings of the tinners, the face of the inaccessible cliffs is maculated with innumerable cavities which penetrate far into the headland. Molybdenite occurs at the junction of the granite with the slate.

The lodes in the neighbourhood of the beacon have been long worked for tin ore, amongst the most celebrated are the old mines of Polberrow and Huel Kitty. There are many tin mines still at work, the most prominent being those of West Kitty, Huel Kitty, and Penhalls. Saint Agnes produces about 700 tons of black tin per annum. This district is remarkable for the disturbances in the relative situation of the rocks, which the intersection of lodes of opposite inclination has caused to take place, and for the dissemination of tin ore through the innumerable small fractures which the rock has suffered. From the exposure of the veins along the cliffs and in elvans, tin ore has been worked from time immemorial. The numerous bold capes which jut out from the coast at right angles to the direction of the groups of lodes and elvans, allow great facilities for the study of their basset edges, and of the heaves occasioned by the crosscourses. The lodes are rather small and quartzose, and slaty matter often predominates. Wood tin was found in the Pye lode at Polberrow, bismuth at Huel Coates, Huel Rock, and West Kitty, and stannite in the latter; at Huel Kind, vivianite and topaz occur.

The tin of Saint Agnes was once supposed to "make" shallow, but the rich tin found in Huel Kitty at 150 fathoms deep has tended to dispel this surmise.

In the sienna and blue slates to the south-west of the beacon, are situated the copper mines of the Huel Towan group; though now idle, 30 years ago they were very productive. There was nothing peculiar in the vein filling, or in the occurence of the ores. Some ores of uranium were met with at Huel Basset, and bitumen at South Huel Towan.

PERRAN MINES.

This parish is distinguished by the varied nature of its mineral productions; its lodes gave existence to the copper mines of Perran Saint George, the rich lead mines of Wheal Golding, the lead and blende mines of the Chivertons, the tin mines of Huels Budnick and Vlow, the iron mines of Duchy Peru, and the silver mine of Huel Mexico. The slates enclosing the copper lodes of Wheal Prudence, Perran Saint George and Wheal Leisure, &c. is usually pale blue, soft and prone to decompose. The copper mines of this group yielded a profit of £250,000. The lead lodes of Penhale and Huel Golding are seen crossing the indentations along the cliffs on which the machinery was perched; they were productive, but have been temporarily abandoned. The tin mines of the Huel Budnick district were not very rich. The Budnick lode, which is associated with an elvan, is remarkable for the heaves which it causes in the cross veins of Penhale and Huel Golding. The lead

lodes of the Chiverton district possess a peculiarity
nowhere else remarked in Cornwall, they have a bearing
approaching that of the tin lodes.   Wheal Chiverton and
West Chiverton yielded enormous masses of galena for
many years, but became exhausted about 10 years ago
after dividing about £230,000.   Much fine blende of
good percentage was raised, and at Great South
Chiverton some blende of a beautiful amber hue was
found in sinking the engine shaft.   The two Chivertons,
like all lead mines in the county, became unproductive
before reaching the 200 fathom level, a depth comparatively shallow.   At Huel Mexico much silver ore was
raised half a century since, in the state of chloride and
sulphide, together with capillary and arborescent silver.

EAST HUEL ROSE DISTRICT.

The east and west lead lodes of the Chiverton mines
prolong themselves eastward into Newlyn, where the
lodes of the Shepherds group were cut in 1816 by some
labourers engaged in draining a marsh,   They were
worked with much profit 40 years ago, and lately have
again been opened.   Besides these lodes, there is a group
of north and south lodes accompanying the Saint Columb
elvan course.   The lode worked under the name of East
Huel Rose was extremely rich thirty years ago, and
gave the adventurers over a quarter of a million profit.
This mine has, within the last three years, been re-started
and very fine pumping machinery has been erected to
drain the workings, which on account of the loose
decomposed character of the slates of this district are wet
and difficult to keep open.

SAINT AUSTELL DISTRICT.

The tract of country in which the copper, tin, and iron mines are situated, lies on the beautifully diversified and wooded slopes that fall from the Hensbarrow granite southward to the sea. The once celebrated Crennis and Pembroke group of copper mines was situated along the cliffs from Par to Charlestown. The lodes produced the usual ores of copper associated with mundic, blende, and spathose iron, and were encased in a thick lamellar rock, pale to deep blue in colour, and often inclined to disintegration. Par Consols was worked to a depth of 250 fathoms, the deeper portion of the mine yielding tin, the profit on working both the copper and tin was £250,000. The lodes of Great Hewas and Polgooth, that accompany the well characterised group of elvan courses that run parallel with the southern edge of the granite, are principally tin producing, although rich courses of copper pyrites have occured occasionally, as at the 90 fathoms level in Polgooth, where the lode was six fathoms wide. In the latter mine, tin was found in the cross-course as well as in the lode. The tin lodes continue westward through Dowgas to the Terras mines. Here to develope a group of tin lodes, efficient stamping machinery has been recently started ; as this is the pioneer mine of the district, the result will be awaited with interest. The gangue is generally composed of quartzose slate or quartz, but with felspar, clay, and schörl when near elvan. The tin lodes of the Huel Eliza group have been rich, and that mine has divided £60,000 on an outlay of £20,000 ; the tin sale in 1882 amounted to nearly 500 tons of black tin.

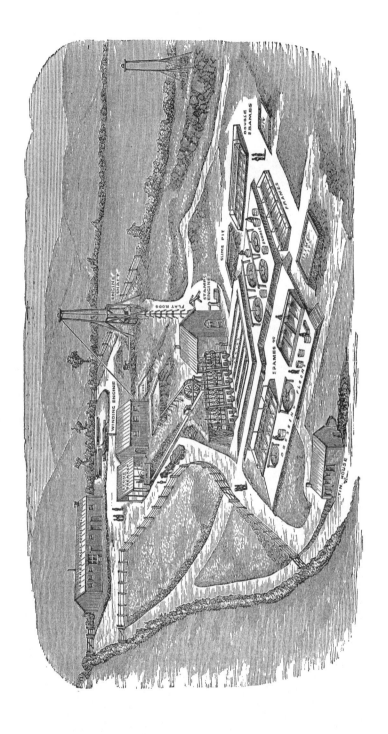

The Fowey Consols copper lodes traverse blue slates often micaceous, or compact felspar rock, which perhaps owe their texture to an underground spur of the granite protruding from the Hensbarrow mass. The gangue consisting chiefly of quartz, slate, chlorite, and spathose iron, encloses mundic, iron, and copper pyrites, vitreous ores, and native copper. The working of the copper lodes gave a profit of £220,000 ; they then changed to tin, but the adventurers not having the patience to sink, the mines were stopped about 20 years since. There is in this mine a perpendicular shaft 300 fathoms deep. The Restormel iron mines to the east of Fowey Consols have sold large quantities of iron ores.*

TIN MINES OF HENSBARROW.

The average height of the granite plateau is about 650 feet. The bleak, ill-cultivated moors, are separated by shallow valleys into rounded elevations which are often crowned by cairns or craggy heights, whilst the slopes are strewed with huge schörlaceous masses *in situ* and with granite boulders and quartz. A dreary, swampy moor, with innumerable excavations and unsightly burrows made in search of stream tin, stretches for miles northward to the isolated granite cones of Belowda and Castle-an-Dinas, The surface over a large extent is much decomposed, and a layer of detritus fills the valley bottoms, rendering them soft and treacherous to the pedestrian. The tin ore in this boss is associated with

---

* The Eselschacht at Küttenberg, Bohemia, is nearly 4000 feet from the surface.

schörl in regular veins, as at the Bunny, Beam, Rocks, and other mines. The veins were usually small, especially when the tin ore was rich, but they seldom continued to any great depth, except at the Beam mine, where one was followed to a depth of ninety fathoms. Much of the tin ground accompanying the veins, owes its stanniferous value to the infiltration of the tin oxide into one ore both sides of the lode, which is often small ; a good example of this class of deposit is exhibited in the Rocks mine, where rich bunches of cassiterite were found so pure that they required scarcely any dressing. At Beam occurs ferric phosphate and wolfram.

These mines were anciently worked as open casts, and some of them were commenced centuries since. The lodes are nearly always in "pot" granite, and the ground has often been worked alternately for tin ore and china clay, so that they better merit the title of stockworks than mines. Not only the schörlaceous veins, but the bands of schörl rock, whose resistance to disintegration has rendered them prominent, enclose a small proportion of tin oxide. Courses of greisen, which are frequent in some localities, also enclose minute quantities. Although tin oxide is so much disseminated throughout the western moiety of Hensbarrow granite, the difficulty of mining operations, and the gradual empoverishment of the lodes in depth, tends to somewhat discourage deep mining. To the east of the Luxulyan gorge, the granite changes its character, becoming dry and hard.

The largest and most interesting stockwork in Cornwall is Carclaze, situated about two miles north of St. Austell. It has the credit of being the oldest excavation in Cornwall, and is said to have been intermittently worked for

500 years. The longest diameter is nearly half-a-mile, the area of the excavation thirteen acres, and the depth 132 feet. Though now worked principally for china clay, the numerous interlacing tin veins of opposite dip enclosed in a schörlaceous granite, were formerly stoped and stamped, and the tin ore separated by washing. At this time there were eight little stamping mills engaged in pulverising the tin stuff, the water and slimes being discharged through a tunnel about 500 yards long, that was also used as a canal, to transport the tin and tin ore to the works on the outside. Only four stamps now exist, and but little black tin is returned, but china clay is collected to the amount of about 5000 tons per annum. The amount of black tin then sold was about 10 tons yearly. The transition rocks are very interesting, and being laid open, can be studied with a facility unparalleled in the county. The granite acquires a stratified structure by the appearance of small schörl veins dipping towards the slate; these grow thicker, and getting mixed with veins of quartz or quartz and felspar, changes to tourmaline schist, which gradually shades off into brownish beds of clayslate.

The Gossmoor is limited on the north by the granitic range of Belowda and Castle-an-Dinas, 765 feet high. Parallel to the range on both sides is a remarkable group of elvan courses running with much persistency east and west. On the south many mines have been opened on a large light brown elvan of which the joints were full of cassiterite, but though Castle-an-Dinas mine was profitable, as a whole the district has not been very productive. To the south-west the copper and tin mines of St. Enoder parish, much worked some forty years since, are now idle;

they were never very rich, but East Basset is said to have declared some dividends.

Some very peculiar deposits of tin ore in the clayslates at Lanivet near Bodmin are worked quarry wise. Mulberry mine is by far the largest excavation, being as deep and nearly as large as Carclaze; the returns in 1882 reached sixty tons of black tin, and Wheal Prosper sold during the same year over twenty tons. The tin ore occurs in the joints and small fissures of the slate like a stockworks, and thus the whole hill has to be trammed away to the stamps and reduced. Reperry mine, besides tin, has many curious antimonial ores. The old Mandlin mine, situate near the eastern border of the Hensbarrow granite, was remarkable for the massive garnet rock which formed the matrix of the lode, and also for the occurrence of the minerals tungstate of lime, magnetic pyrites, wolfram and others.

To the north of St. Breock Downs, in the vicinity of Padstow, are a few north and south lead lodes which up to the present have led to no great mining operations. In the Delabole district are a few unimportant veins which enclose galena in small quantity. At Trugoe in St. Columb parish some copper ore, bismuth, and cobalt blume have been met with, and bournonite in St. Merryn.

CARADON DISTRICT.

The barren aspect of the Bodmin Moor is reflected by its rocks, which are very destitute of metallic ores, and it is only on the southern fringe of the granite that copper and tin ores abound. At Roughtor, east of Camelford,

a large sum was expended to discover whether the tin
veins in granite improved with depth, but the failure was
complete. At Blisland where there is a well marked,
though very granitic group of elvans, no lodes of any
promise have yet been noticed, but no explorations of
importance have been made.

The mines around the Caradon Hill—1208 feet high—
were originated by some miners driving an adit in 1836,
but though comparatively modern, after a brilliant existence the first fruits of the district have been gathered,
and the mines once so numerous and prosperous are now
mostly stopped. South Caradon, the first mine opened,
yielded 9% ore, and gave for many years handsome dividends, the total profit having been £380,000. The copper
group extends eastward through East Caradon to Glasgow
Caradon both very profitable mines. To the north is the
Phœnix group of tin veins, where owing to the projection
of granite ridges, and the faulting of the lodes, the hanging wall is slate, whilst the foot wall is often granite. The
matrix of the tin ore is composed of quartz, chlorite and
earthy iron ore. Adjacent the surface, copper pyrites,
and malachite are found. Nearly all the lodes dip steeply
towards the granite, and have an average width of rather
more than three feet. At Gonamena tin ore is found in
a manner somewhat resembling Carclaze, the excavation
is a third of a mile long, and occupies a dozen acres, but
the depth is only fifty feet.

To the west, the lodes are principally tin producing, and
continue with a group of elvans through St. Neot to
Warleggan. Though the mines have only been worked
in a partial and desultory way, there is ample evidence
that good tin lodes, which merit exploration, exist. At

a mine called Tin Hill a large quantity of stream tin was
obtained from a remarkable deposit of gravel and boulders
deposited beneath cliffy granite. Some elvan courses
have been worked for tin with moderate success in this
district.

TRELAWNY LEAD DISTRICT.

Three miles south of Caradon Hill is an interesting
group of mines that have produced very large quantities
of argentiferous galena. Wheal Ludcott created great
excitement in the mining world in 1861, on account of the
discovery of native silver, both capillary and arboriform,
in an east and west cross course. The principal mines
were on the Trelawny lode, and they all returned large
profits on the capital employed. The lode was accompanied by a capel of brecciated structure, and the galena
—which contained about fifty ounces of silver to the ton
—-was enclosed in a gangue of quartz, with some fluor,
blende, copper and iron pyrites.

South of St. Neot, close to the road from Bodmin to
Liskeard, is the East Jane group of lead lodes. In the
clay slates of Lanreath and Duloe, five miles south-west
of Liskeard, the Herodsfoot mine has yielded much silver
lead ore, and beautiful specimens of bournonite (sulphide
of lead, copper and antimony), and sulphuret of antimony
which is found in quantity in no other Cornish mine.
The veinstuff is quartz, fluor and calcspath. The lode
does not now produce so much lead, in 1882 only 362
tons of galena containing 8130 ounces of silver were sold.
At North Herodsfoot during the same year 63 tons of
silver lead were sold.

CALLINGTON MINING DISTRICT.

The mines included under this name are those encircling the granite domes of Kit Hill and Kingston Downs west of the Tamar. Just north of Callington is the copper, tin and lead mine of Redmoor, situate on the western flank of Kit Hill, in a thick lamellar blue slate of silky lustre. The copper and tin are found in east and west lodes, and the lead in the crosscourse; this is the only mine in the county where veins crossing each other have both been productive of metallic mineral in quantity. The gangue of the former was quartz, slaty clay, and where near the elvan felspar clay; in the latter quartz, slaty clay, iron pyrites, and some ferric carbonate formed the matrix. Redmoor mine sold in 1878 cupreous precipitate, worth 114 ounces of silver to the ton. Ores of silver were found in Wheal Dudley and Wheal Vincent. The cone of Kit Hill, through which pass numerous small veins, is being cross-cutted by a deep adit level from the north side. The tin ore is associated with wolfram.

The granitic mass of Gunnislake, though scarcely two miles long, is lofty and the declivities singularly abrupt. The beautiful valley of the Tamar, with its orchards and gardens, winds around the eastern margin through a deep and romantic gorge. Hingston Down mine is situate near the summit, and at the foot of the northern slope are the tin and copper mines of New Great Consols, Lamerhoo, Devon Great Consols and others, whilst the silver mines, Prince of Wales, Huel Brothers, Harrowbarrow and Wheal Newton, together with the tin mines of Drakewalls, dot the wooded foot hills of the southern slope. Several mines in the Callington district produce

argentiferous copper ores in a quartzose gangue with
mundic and mispickel. The ores of New Great Consols
contain much arsenic, silver, copper, and tin, and from
these were reduced arsenic, argentiferous precipitate, and
black tin. The precipitate sold in 1875 fetched £3000.
The Cornwall Chemical Company erected large arsenic
and precipitating works at Greenhill, in Calstock, to treat
the ores from their mines at Holmbush, Kelly Bray,
Wheal Newton, and Greenhill. They put a splendid
reduction plant that would have ensured success, if capital had not suddenly fallen short, and led to an abrupt
suspension. From these mines large quantities of pyrites
were sold, and the silver ore produced at Wheal Newton
contained 26,800 ounces of silver. The only copper
mine in operation on the Cornish side is Gunnislake
(Clitters), which returned in 1882 copper ore to the value
of £14,000. The group of stanniferous lodes at Drakewalls were worked anciently, though at present less than
50 tons annually are returned. It was at this mine that
the wolfram which contaminated the black tin, was converted into a marketable commodity as tungslate of soda.
Molybdenite occurs in this mine. The Huel Brothers
parallel group of veins yielded much silver, and the gangues raised from the various mines—Harrowbarrrow,
Silver Hill, Prince of Wales and others—are considered
to possess an average value of several ounces of silver to
the ton. The Huel Brothers lode which traversed a
decomposed pale blue slate, produced red, vitreous, and
black silver ores with native silver, associated with
galena, spathose iron, blende, iron and copper pyrites,
enclosed in quartz, slaty and felspar clay.

TAVISTOCK MINES.

The river Tamar, which separates the counties of Devon and Cornwall, has its source within three miles of the north coast, in the neighbourhood of Morwinstow. The country from Tavistock southward is much diversified by the meanderings of the Tamar with its multitudinous tributaries, and the deep valleys separated by narrow ridges, picturesquely variegated by woods, orchards, and gardens and sheets of water, shut in by softly outlined hills, afford an alternation of landscape and marine scenery delightful to the eye and pleasing to the imagination. Notwithstanding that the mines of this district are in the county of Devon, they are so closely connected mineralogically as well as geologically, that the description of the mining fields of East Cornwall would be incomplete without some notice of the rich copper and lead mines grouped around the prettily situated town of Tavistock.

A group of rich copper lodes extend from Gunnislake granite across the thick lamellar blue slates of the Tamar valley eastward to the Dartmoor range, but are most productive near the former. A little to the north of the eruptive boss on the same group of veins as New Great Consols, is the celebrated Devon Great Consols, the richest copper mine in the two counties, which in twenty years declared more than a million sterling in dividends. Eastward from the great crosscourse, the lode becomes poorer, and consequently, to treat the ores successfully, dressing machinery had to be erected. The company possess the most complete separating plant in the west. In spite of the low produce of the copper ore raised, the sales of copper ore in 1882 aggregated 12,000 tons, besides

2760 tons of refined arsenic. Bedford United returned
£54,000 in dividends during the earlier working for copper, and has lately again become profitable. Wheal Crebor,
another copper mine, opened on the lodes of the same
group, sold £10,000 of copper ore in 1882. To the north
of Tavistock are the Wheal Friendship copper mines, commenced in 1798, which gave a profit of £300,000; a small
quantity of tin was also obtained.

The crosscourses running north from the Friendship
group were worked at Huel Betsy, and near Lidford for
galena.   To the south of Drakewalls mine are the
meridional lodes of Beerferris and Beer Alston, which
crossing the Tamar twice, were worked under the bed of
that river.   The mines were lost, in consequence of the
lode having been stoped up close to the bed of the river,
which broke into the workings suddenly on a Sunday
morning; had the accident happened on a working day
300 miners would have lost their lives. These mines
were remarkably profitable, the galena produced being
highly argentiferous. Several rare and beautiful phosphates, sulphates and carbonates of lead, and carbonates
of copper were taken from the lodes. The company
smelted the lead, and extracted the silver at their works
on the Tamar at Ware Quay.

In the Devonian rocks which rest on the granite from
Horrabridge around to Buckfastleigh are some copper
and tin mines. At Bottle Hill tin has been largely
raised, and the copper mines near Ashburton yielded very
rich ores. The carboniferous slates which skirt the northern margin of the Dartmoor granite are non-productive
unless in highly metamorphic rocks at the junction. Thus
at Belstone, where garnet rock and greenstone are inter-

colated with slates, some copper ore has been obtained, and in 1882 a few hundred pounds worth were sold. Numerous mines of manganese are found in the carboniferous beds to the north of Launceston, and a considerable tonnage sold. Manganese was also raised in Calstock— near the Tamar.

CONCLUDING REMARKS.

No one who has studied attentively the metalliferous strata of Cornwall, and the mode in which the metallic ores occur in the fissures associated with them, can feel convinced that the rich tin and copper lodes, which at comparatively shallow depths, have been " worked poor" are exhausted, or that the analagous rocks mantling around the granite bosses may not enclose undiscovered lodes, maugre the thorough rummaging of the old tinners. The Caradon group of copper lodes discovered in the middle of the present century, is a remarkable instance of unsuspected mineral wealth, and one which finally dispelled the conviction of " Cousin Jack " that no copper could exist east of Truro Bridge.   The richest mines must come to an end when the courses of ore are stoped away and no provision is made to pierce the poor bar of ground which the theory of vein formation makes manifest, must as a general rule intervene between two deposits.

A knowledge of where *not* to explore is important, and —speaking for Cornwall—great expenditure would be injudicious, whenever there is an absence of metamorphosed rock and massive felspathic or thick lamellar strata of variable. density.   Lodes found in such rocks, carrying true gossany backs, may be followed with confidence, though the presence of oxidised iron is not absolutely indispensable to the occurrence of rich courses of ore. A chain of favourable events are requisite for the development of ore bearing lodes, and as these cannot be supposed to have happened everywhere, there are many wide tracts where the metamorphic rocks do not possess all the characteristics which are distinctive of good metalliferous districts.

Careful examination of the transition rocks around the granite by intelligent miners, with some capital discreetly applied to drive crosscut adits where auspicious conditions prevail, might, and very probably would, lead to discoveries equalling that of South Caradon, because many of the richest copper mines in the county have been discovered by accident. Less than half a century since, before promoters and mine brokers existed, small proprietors, farmers, and even artisans would combine to cut some lodes by an adit or shaft, and these ventures often led to important results. Nowithstanding that the lodes in abandoned districts such as Gwennap would at greater depth dislose rich masses of ore, the expense of re-opening and sinking through long bars of poor ground in search of them opposes almost insuperable obstacles while metals remain in the present depressed condition, but the future, possibly not so very remote, may yet see these mines working and producing large quantities of tin.

The great fall in the value of the metals produced by Cornwall during the last few years, has led to such a serious decline in the mining industry, that thousands of Cornishmen, have been compelled to leave the county, and may now be met with in all the mining fields of the world. In 1836 no less than 30,000 persons were engaged in Cornish mines, it is doubtful if half that number find employment in that pursuit to-day.

To alleviate to some extent the present dangerous state of mining industry, *en attendant* a rise in prices, the introduction of machinery to expedite operations is of much importance. The loss in dressing tin ores is believed to be about one fifth of the tin extracted from the stopes, and if some means could be devised to stop even a third of this loss it would take mines off the calling list. Dolcoath and other mines are fast approaching a depth of half a mile, so that first class machinery for hauling the ores to surface, and convenience for the circulation of the mines has grown absolutely indispensable. To meet successfully the present crisis, it is essential to alter the mining customs, and to introduce a proper mining code assimilating as near as possible to those of Continental countries, whose governments have been watchful in byegone times to prevent the owners of the soil from usurping the national treasures buried beneath the soil.

The dependance of the west on the industries associated with mining, is a subject too well known and appreciated by landowners, farmers, and workmen, to require any observation here, but it may be remarked that the stoppage of the mines, would result in the semi-depopulation of the western towns, and a return of the inhabitants to their ancient condition of agriculturists and fishermen.

CHAPTER V.

GEOLOGICAL ECONOMICS.

A bare description of the rocks and minerals of Cornwall, and the localities in which they occur, would be perhaps unsatisfactory to the reader unaccompanied by some particulars of the processes by which the useful products are prepared for the market, and the uses to which they are applied in manufactures. The space alloted to this geological sketch, will allow of only a rapid review of the subjects of interest connected with each metal; but the preparation of porcelain clay, on account of its interest and importance (400,000 tons being annually exported), and its fabrication into such numerous articles of domestic utility, will be treated with a considerable amount of detail.

THE CHINA CLAY AND CHINA STONE PRODUCTS.

The Granite Rocks of the West of England have long yielded materials for the use of the potter. These consist chiefly of a fine white refractory clay, called indifferently kaolin, porcelain clay, china clay, or Cornish clay; and a white vitrifiable variety of partially decomposed granite, known as porcelain stone, china stone,* or Cornish stone.

The rise, progress, and present condition of so important an industry cannot fail to be interesting to many. No very complete account of this industry was ever published before the year 1881, when Mr. David Cock, of Roche, published "A Treatise, Technical and Pratical, on the nature, production, and uses of China Clay." Mr. J. H. Collins, F.G.S, in his able and interesting treatise on the Hensbarrow Granite District, says that the earliest direct mention of the clay-works is that by Dr. Pryce, who stated that china clay, was in 1778, prepared for the potters in the parishes of Breage and St. Stephens, and packed in casks for exportation. The trade was, however, many years old when Pryce wrote, as Mr. William Cookworthy, in Devon, had established its value more than 20 years before, and had used it under a patent in his own pottery which was established in Plymouth in 1733. It is not clear whence he obtained his first clays, but in a pamphlet published in 1853, a letter from Cookworthy to a friend is inserted, which states that an American brought him specimens of **Kaolin** found in Virginia, and also specimens of porcelain made therefrom. This material, be observed, could be imported for £13 per ton.

The American and his specimens set Mr. Cookworthy on the *qui vive* and very soon he found a stone resembling

---

* No doubt the name of " china " clay and stone is derived from the fact that the porcelain goods were first made in china ; and which, 60 years ago, were sold at very high prices.

L

that shewn him, in St. Stephens, probably about the
year 1755. Three years later he found a similar
material in the parish of Breage.

It is probable that the Cornish clay was known
to him, and that he had already used it to some extent,
keeping the fact secret, after the manner of the times.
At any rate he patented the use of these materials in
conjunction with Lord Camelford, in 1768. Dr. Borlase,
who wrote in 1758, does not say that Cornish clay was
then employed, but he states that many suitable clays
may be found in Cornwall, and especially mentions
those in Towednack, Tregoning Hill, and St. Enoder.
He also states that he had himself made experiments
with the clay of Towednack, and that Mr. Cookworthy
had tried that of Tregoning Hill. This latter locality,
therefore, was probably the birthplace of the Cornish
china clay trade, as St. Stephens was of the trade in china
stone.

Dr. Thomson, who visited Cornwall in 1813, briefly
referred to the condition of the china clay districts as he
saw them, in the " Annals of Philosophy " for that year.
Dr. Fitton, in an admirable paper which he contributed
to the same journal a few months later, describes what
he saw when he visited Cornwall 6 years previous to Dr.
Thompson's visit. He stated that were then seven works
in operation in the parishes of St. Dennis and St.
Stephens, the largest of which produced 300 tons per
annum.

The next notice with which I am acquainted, says Mr.
Collins, occurs in a short paper contributed to the Royal
Geological Society of Cornwall, and printed in the year
1818. The writer, Dr. Paris, gives the quantities of

china clay and stone shipped from the port of Charlestown in 1816—1817; and he remarks that the amount of royalty, or dues, paid to Lord Grenville was £700 per annum! The West of England Company alone now pay to his representative a minimum rent of £7000 per annum. The total quantity shipped at Charlestown was under 4000 tons, and it is not likely that more than a few hundred tons were shipped from any other port at that time.

MODE OF OCCURRENCE.

In all the masses of eruptive rock dotting the county, some portions are productive of a fine refractory clay which is used in the highest branches of ceramic art. The bulk of the clays and china stone produced, comes from the granite lying to the south and west of Hensbarrow Beacon, only small quantities being furnished by Tregoning Hill, and some small works in Towednack, and Blisland.

Mr. J. H. Collins, who has closely observed the decomposed granite north of St. Austell, considers china clay in its natural state to be simply a granite of white or pale smoky quartz, white mica, and white felspar, in which the latter is partly or completely changed to kaolin. This rock is constantly associated with parallel groups of quartzose or schörlaceous veins which include also tin oxide; indeed, many of the kaolin deposits continue in the direction of veins for as much as a mile in length, while their breadth may be but a few feet. A very fine vein may be sufficient to influence the softening of the granite to a great width and depth. The decomposition of the felspar, which is a silicate of alumina and potash, permitting the latter to be carried off in solution, leaves a hydrous silicate of alumina called kaolin or china clay. China stone is a mixture of quartz and more or less softened felspar which is quarried and exported to the potteries without prapara-tion. Lately the West of England Company have had mills erected near St. Blazey to grind the stone to powder.

The natural clay rock is always covered with a thick layer of stones, sand, or impure and discoloured clay, known as "overburden." It varies from three to forty feet in thickness ; and it must, of course, be removed before the clay can be wrought. The decomposed granite is found at all elevations except the very highest points of the districts, which are always composed of hard rocks—and its situation is usually indicated to the practised eye by a depression of the surface. These depressions are not observed in the case of china stone. The natural clay rock, being a decomposed granite, consists of kaolin, irregular crystals of quartz, and flakes of mica, with sometimes a little schörl.

MODE OF WORKING.

At the time of Dr. Fitton's visit, 70 years ago, Tre-thosa was one of the largest works, but it produced only about 300 tons per annum ; now there are many works producing twenty times as much. The old fashioned system, somewhat modified in detail, but the same in principle, still survives in a few places. The following descriptions apply with more or less accuracy to a majority of the larger works of the present day, turning out from

2500 to 8000 tons of clay each, yearly. Two somewhat
different methods are employed, according to the situation
of the "bed" of clay, in relation to the surface contour
of the immediate neighbourhood. The most general case
is that in which the clay has to be raised from a veritable
pit, the bottom of which is lower than the ground in the
immediate neighbourhood on all sides.

The exact situation of the clay is first determined by
systematic "pitting" to a depth of several feet or
fathoms, or occasionally by boring. A shaft is then sunk
either in the clay itself or preferably in the granite close
to the clay. From the bottom of this shaft a level is
driven out under that part of the clay which it is intended
to work first, and a "rise" is put up to the surface,
which should by this time be partially cleared of its overburden. The common depth for such a shaft is about 10
or 12 fathoms. As soon as the rise is completed to surface, a "button hole" launder is placed in it, and the
remainder of the rise is filled up with clay. In the
meantime a column of pumps has been placed in the
shaft—say from 10 inches to 12 inches in diameter, and
an engine erected to work them, unless water power is
attainable. Owing to the low prices obtained for some of
the clays, steam power in many cases is too expensive ;
but it will do in such a clay work as Rosemellyn, near
Bugle, which yielded clay of superior quality.

For water, many works are almost entirely dependent
upon that met with in sinking the shaft and driving
levels ; but of course this may be, and is, increased in
some places by storing rain water in reservoirs, and by
making use of such small streams as may be available. A
small constant supply is sufficient even for a large work,

as it is used over and over again ; the clay in suspension
being pumped up by the engine into reservoirs where the
clay is deposited and the water becomes almost clear. The
work commences around the upper end of the " button
hole" launder, by running a stream of water over the
exposed clay and breaking up the "stope," with picks.
A large quantity of sand is constantly produced, and as
constantly shovelled out of the way into wagons and removed ; while the water, holding the clay and fine impurities in suspension, runs down the launder, along the
level, and into the bottom of the shaft, whence it is
pumped up by the engine.

As the excavation becomes larger and deeper, more
overburden is removed, and the upper portions of the
launder are taken away, until at last the stopes reach the
level, when the launder is, of course, no longer required.

At first the sand is thrown out by one or two "throws"
but very soon it becomes necessary to put in an inclined
tramway for raising the sand in wagons, and this is
worked by a horse-whim, or by winding gear attached to
the engine or water-wheel. As there are from 3 to 8 tons of
sand produced in getting each ton of clay, of course its
removal in the cheapest possible manner is a matter of
great importance. Any veins or lodes of stone, or dislocated portions of clay, are raised from the " bottoms " in
the same way as the sand.

The stream of water holding clay, fine sand, and mica,
in suspension, is, in well-arranged works, lifted at once
high enough to allow of all subsequent operations being
carried out by the aid of gravity. The stream is first led
into one or two long channels, the sides of which are
built of rough stone. In these channels, called " drags,"

the current suffers a partial check, and the fine sand, and rougher particles of mica are deposited. From these drags the stream passes into other channels much resembling them, but of greater number, so as to divide the stream still further. The second series of channels, known as "micas," is often built of wood, but sometimes of stone. They differ in no respect, essentially, from the "drag," but are more carefully constructed, and better looked after ; and as the stream is greatly divided, and is very gentle, the fine mica is deposited in them. The "micas" are often about eleven inches wide, ten or dozen in number, and one hundred feet or more in length. Provision is made, by the underground channels and plug holes, for the periodical cleansing of the drags and micas. This may have to be done twice a day, but generally only once. The deposit of the drags is worthless at present, and is always thrown away, but that from the "micas" is often saved, and sold as inferior or "mica" clay. The refined stream of clay then passes on to the "pits," which are often made circular, 30 to 40 feet diameter, and 7 to 10 feet deep.

These pits are built of rough masonry, and they have an outlet at the bottom opposite the point at which the stream of clay water is admitted. This outlet is stopped by a "hatch," or else by a plug, which is kept closed until the pit is full of clay. In each outlet, however, is fixed an upright launder about 4 or 5 inches square, provided with "pin holes" and wooden pins set close together, or near each other. As the stream of clay enters on one side it is continually depositing its burden, and the water runs off, nearly clear, from the pin holes higher and higher as the clay rises in the pit.

The affluent water is conducted directly to small storage reservoirs, and thence over the clay stopes, whence it does its work over again. It may here be mentioned, that when the stream of clay water enters the pits, it contains from $1\frac{1}{2}$ to 3 per cent of clay, and what is called a good washing stream will carry about 1 ton of clay per hour.

When the pit is full the "hatch" is drawn, and the clay is "landed" into the tank. The upper portion is sufficiently fluid to run in of itself, but that near the bottom has to be helped out by men using "shivers" of wood or iron, which resemble large hoes, and by a small stream of water. The tanks are commonly, but not always, rectangular, built of stone, and paved with stone at bottom, often 60′ by 30′ by 6,′ or even larger. Once in the tank, the clay is left to settle until it has the consistency of cream cheese, the water being drawn off from time to time, when it is ready to be trammed into the dry.

The "dry" is a large building erected contiguous to the tanks. It is always composed of two parts, the dry proper and the "linhay" or shed. The floor, or "pan," of the dry is composed of fire-clay tiles, 18 inches square, 5 or 6 inches thick at the fire end, and gradually thinning off to 2 or $2\frac{1}{2}$ at the stack end. The flues are built of fire brick about 14 inches wide, 2 feet deep at the fire end, and 9 inches deep at the stack end. Each flue should be supplied with a damper. The furnaces are built in and arched over with the best fire brick; the fire bars run longitudinally, and are about 6 feet long. The grate surface is about 2 feet 6 inches wide in front, and 4 feet 6 inches to 6 feet at back, according as each furnace supplies three or four flues.

The clay is brought in from the tanks in tram wagons, holding about half a ton, tipped on to the tiles, and spread in a layer from 9 inches thick at the flue end to 6 inches thick at the stack end. The fire end is loaded and cleared every day ; the other end perhaps twice or thrice a week, according to the length of the dry, thickness of tiles, perfection of draught, &c. An average size for a first-class dry is about 15 feet wide, and 120 feet long, but some have been constructed considerably larger than this.

The pan of the dry should be 6 or 8 feet above the linhay whenever possible, so as to afford storage space for the dry clay without expending labour in piling. The tiles should be as porous as possible, for very much more water passes through the tiles and into the flues than is driven upwards in the state of steam.

When the clay in the dry is nearly free from moisture, it is cut into squares of about 9 inches ; and when it is perfectly free from moisture it is removed and placed in the linhay ready for transit to the railway station or ship. In some cases it has to be carted to the station at a cost of about 2/6 per ton. This is a charge which seriously diminishes the producers' profit. In several places the dries are close by the Minerals railway as is the case at Fal Valley, Bugle, Burngullow, Rosemellyn, &c.

USES OF CHINA CLAY.

The first use to which china clay was applied as already stated, was the manufacture of porcelain, and this is still popularly believed to be its sole use. This, however, is

by no means the case—probably little more than one-
third of the produce is so applied. Large quantities are
used by bleachers for filling up the pores of calicoes
as a dressing ; and still larger quantities are used by
paper makers to give " body " and weight to their paper,
especially printing papers. The manufacture of alum,
sulphate of alumina, and ultramarine uses up large quantities annually. Small quantities are used by photogra-
phers, manufacturing chemists, and colour makers, for a
great variety of purposes ; and, if reports are to be
believed, it has been used in the adulteration of flour, and
of artificial manures.

COST OF PRODUCTION.

A work capable of producing say 4000 tons of china
clay yearly will cost from £2500 to £5000. To get the
clay in the shed ready for the market, will cost about 9/-
per ton, of which 2/6 must be expended in fuel for pumping and drying, 1/- in removing over burden, 1/- in
removing sand, and 1/- for management and office expen-
ses, leaving 3/6 as the net labour cost of washing and
drying a ton of clay.

To the 9/- net cost of clay, must be added 3/- for royal-
ties, 4/- for transit and shipment and 1/- for commission,
bad debts, and sundries ; making the average actual cost
amount to 17/-. Some favourably situated works can,
no doubt, save two, or even 3/- of this amount ; in others
the cost may amount to 20/- or even 22/-.

As to the selling price ; this varies more widely than
the cost of production ; at present, prices are very low ;

ranging from 14/- to 35/- f.o.b.   Clays sold at the lower
rates must be unremunerative.

In the year 1809, the produce of Cornwall and Devon
was 1757 tons of china clay, and 1162 of china stone ; in
1882 the produce was 270,910 tons of clay, and of stone
35,737 tons, showing an enormous increase of production,
but owing to the competition amongst the producers, the
average price per ton was ruinously low, being only 15/-
per ton.   Since 1853 more than five million tons of china
clay and stone have been exported.

Ochre.—Some small quantities of ochre and umber
have been produced.   In the year 1857 twenty-five tons
of umber were sold from a small excavation near Indian
Queens.   In the Terras mining district some adits which
drain the mines, become partially choked by the oxide of
iron deposited from the super saturated waters.   Workmen go in and agitate the bottom with rakes, and the
ochre is carried out in suspension and deposited in " catch
pits ; where, after the moisture has evaporated, it is sent
away to be used in the preparation of iron paints.   The
low price—ten shillings per ton—of the ochre discourages
the production, and the sales are insignificant ; some years
not a hundred tons are sold, but when the adits require
clearing, the sales increase to many hundreds of tons.
During the past ten years about 700 tons have been sold.

Fluor Spar.—This mineral was raised in some
quantity from Huel Gorland, where it formed the veinstone in a lode, and was sent to the Swansea copper
smelters as a flux.   It was sold on the mine at 20/- to
10/- per ton.   Very little is now found.

Saint Agnes Clay.—The remarkable deposit of
sand and clay near the foot of St. Agnes. Beacon is

worked for economic purposes. The light siliceous sand
of great purity is used to mix with china clay in the
manufacture of Cornish crucibles, and the soft plastic
clay is sold to the miners for the purpose of sticking the
candles to their hatcaps, and some is exported to Wales
for furnace linings.

The clay sold in 1882 amounted to 3400 tons, and
realised a mean price of 9/6 per ton.

The following table, gives the number of tons of china
clay, china stone, and St. Agnes clay sold from 1853 to
the present.

Statistical Table.

	CHINA CLAY.	CHINA STONE.	ST. AGNES CLAY.
1838	20,784	7344	
1853	17,000	9000	
1854	18,742	8246	
1855	60,188	19,961	
1856	65,510	7800	
1857			
1858	65,600	21,983	2052
1859	61,470	20,750	
1860	63,250	21,500	1748
1861	60,750	19,700	1894
1862	61,550	19,250	2208
1863	92,500	23,750	1900
1864	95,730	21,570	1450
1865	97,750	25,500	2016
1866	105,000	35,000	2024
1867	127,000	33,500	1816
1868	100,000	29,000	1479
1869	105,000	28,500	1375
1870	110,520	32,500	1189
1871	125,000	33,000	1601
1872	141,000	48,000	2022

Statistical Table—*continued*.

	China Clay.	China Stone.	St. Agnes Clay.
1873	153,000	45,000	2233
1874	150,500	42,500	1818
1875	108,250	38,000	1975
1876	105,275	34,500	390
1877	200,345	39,500	905
1878	185,203	41,250	1169
1879	273,862	38,142	682
1880	278,572	34,870	735
1881	241,658	30,479	825
1882	270,910	35,737	3400

CHAPTER VI.

TIN.

History.—This metal is believed to have been discovered originally by the inhabitants digging for peat. The time when tin was first worked is lost in obscurity, but during all historic ages, Cornwall has been famed for its stanniferous wealth. Before the recent discovery of cassiterite in Australia, the only places from which tin was obtained in quantity was from the mines in Germany—said to have been discovered by a Cornishman in 1240, and from the tin islands in the Java sea, where alluvial tin ore was first collected about the year 1710. Thirty centuries ago Phœnician vessels sailed through the Pillars of Herucles in search of the then rare metal, tin, which the mines of Spain had failed to produce in quantity sufficient to satisfy the demands of an expanding civilization. The commerce for tin between the ancient inhabitants of Cornwall and the Phœnicians is so old, that it was in active existence during, and even before the epoch of the Grecian Myths. The secret of the Cassiterides was so well kept, that it was not until 400 B.C. that some Greeks who had settled at Marseilles sent an expedition in search of them, they were discovered; and the great Mediterranean port shared with the

Phœnicians the advantages of the traffic at Iktis. The demand of the eastern nations must have had an exhausting effect on the deposits of alluvial tin, as we learn that in the time of Diodorus Siculus, the tinners had commenced to barrow in the backs of the lodes.

Many interesting remains of the Phœnician era have from time to time been found, some of which may be studied at the Truro Museum. Double pigs of tin, cast for transportation on mule back, have been picked out of the sea in Carrick Road; golden cups and collars, made out of the gold obtained from the stream works, with Greek and Roman coins, have frequently been discovered. After the destruction of Carthage, the tin trade fell into the hands of the Romans; and they seem also to have worked the tin mines, as their coins have been found in and near them. It is not probable that the Latins had a great ascendancy in Cornwall, though they may have founded a trading port at Tregony; in those days the nearest tidal port to the great tin streams of the Hensbarrow granite.

Tin Streams.—The tin layers which rested in the depressions of the granite hills, and the tin ore carried down into the valleys below, were the result of ages of denuding action. In the St. Austell district, the lodes containing tin being harder, stand out of the ground like stone walls. It is not improbable, that the granite was so much higher than at present, on its emergence from the sea, as to allow of a grade steep enough for the torrents to bear along the tin stone deposited at Carnon, Pentewan, and other places far from the tin bearing rocks. During thousands of years the " old men " were engaged in washing the tin ore which the degradation of the mountains of granite had freed, and the labour lavished is strongly attested by the countless burrows and pits which still disfigure the " bottoms" of the streams flowing from the granite domes towards the sea.

About the time of the Christian era, vein mining was carried on all through the county, and that the industry was well sustained, and the mines worked as far as the appliances of the age would admit, will be readily credited by those whose avocations have enabled them to learn that scarcely a tin lode exists in Cornwall which has not been opened by the pits and adits of the ubiquitous tinner. In the reign of King John the production of tin in Cornwall was so small that the Bishop of Exeter received in lieu of his tithe only £6 : 13 : 4. Tin mining was much favoured by Richard, Duke of Cornwall, who, in the 16th century granted the tinners a charter which has since developed into the Stannaries. The method of washing the tin streams was simple, and the rude utensils employed were usually fashioned out of hard wood. It is also recorded that in Carew's time, tinstuff from the veins was washed on turf, on the same principle as gold ore is washed on blankets in Brazil. Tin Bounds have been worked within the present century, but they are now obsolete, nor are many stream works in active existence ; as they have been streamed again and again, they are now almost exhausted. The largest deposits of stream tin were found in the Goss (Tregoss) moor, a marshy tract containing about 10 square miles, situate between Belovely and St. Dennis beacons. The stanniferous deposit is not completely exhausted in this tract, but it can only be profitably worked when the price of tin is exceptionally high.

Greenstone Rocks. ── BOTALLACK MINE.

Pub.d by H. Besley, Directory Office, South St., Exeter.

The stream tin generally reposes directly on the rock, and is often in a remarkably clean state, with but little admixture of gravel. To account for this fact, some geologists have even imagined that a vast rush of water passed southward over the granite hills. Most of the best streams run to the south ; but this is owing to the rivers that fall into the English Channel taking their rise north of the principal ranges of granite. The tin *debris* resting on the granite is usually concealed under peat, sand, or mud, with or without stones and shells; and underlaid by wood, leaves, and vegetable matter; the whole varying from five to twenty-five feet in depth. Near the sea at Pentuan and Carnon, this is mingled with miner's tools, oyster shells, animal remains, and even human skulls. The tin stratum at the Happy Union stream works, covered by 40 feet of gravel, ferruginous clay, black peat, sea sand, and shelly mud, consisted of fragments of clay slate, quartz, iron ore, &c., but no granite. The tin having been washed and rolled along the river bed for ages, has acquired a rounded form, and that degree of purity which gives stream tin so high a value in the market. The size of the tin stone varies from pieces a few pounds in weight to sand, and associated with it in all the large works were nuggets and prills of gold. The tin stratum in Carnon valley continued into the Restronguet creek, and was worked in 1700, at low water, by hundreds of men, women, and children. About 1865 the creek was worked opposite Devoran by means of a shaft surrounded by a mound. Wheal Caudle in the Helston Looe Pool was worked 30 years ago beneath about 30 feet of sand and mud. It is believed that Hensbarrow granite has yielded more

M

stream tin than the aggregated produce of all the other
granite elevations.

Tin Dressing.—The ore from some lodes (as at
Wheal Lovell) is so rich as to require little preparation
for the smelter ; but practically all the tinstone has to be
stamped and dressed. The tinstuff is broken in the
levels and from the stopes in the mines, and put into the
kibble or skip without any sorting whatever, is hauled to
grass, discharged into the waggon by the "lander," and
trammed into the hoppers whence it is fed automatically
through the "pass" into the stamps "cofer." Here the
tinstuff is reduced to a fine grain, the size of which,
depending on the class of tin, is regulated by the stamps
grates fixed around the cofer-box. Should some of the
tin ore vary widely in quality, or contain much copper
pyrites, wolfram, &c., it is dressed apart.

In front of the stamps are the various strips, tyes, and
buddles, by which the black tin is separated from the
sand. The slimes carried away in suspension by the
water are caught in special pits, trunked and framed.
The last impurities are eliminated by "tossing." Ores
containing arsenic are roasted after concentration, and
re-washed. Although the intricate maze of tin dressing
apparatus, often jumbled together with scant order, may
appear to a visitor a puzzle not be unravelled, in reality
the truly different operations are very few, and most of
the work is a repetition, having for object the rejection of
all foreign matter, so that the tin ore may be sold as free
as possible from any substances which would lower its
price at the smelters. The view of New Terras mine,
near St. Austell, will give the reader some idea of the
arrangement of the dressing machinery. As many mines

produce tinstuff containing only one or two per cent of
tin, the numerous washings which it undergoes result in
a considerable loss of fine tin ore, which certainly equals,
if it does not exceed, a fifth of the tin that the vein stone
originally enclosed. This loss is due in some measure to
the difficulty of getting that portion of the ores already
stamped to the size required through the gratings, which
is thus reduced to slimes that carry away with them the
floured tin oxide. Possibly for many ores, a change to
dry grinding, and jigging, somewhat similiar to the
method of treating copper ores, would result in an
improved yield. The slimes retain the tin so tenaciously
that the tailings from the mines in Camborne pass over
many hundreds of buddles and frames in its course of
several miles to the sea, which it enters still charged with
an important proportion.

Smelting.—The cassiterite, freed as much as possible
from foreign matter, is transported to the smelting-works,
and sold at the standard which the smelters themselves
fix. Here, if the ores are contaminated by iron or
copper they are digested in acid. Anciently the tin ore
was smelted in furnaces scoped out of the ground, by
mixing it with charcoal, and using as blast a rude
bellows. In the last century smelting took place in
blast furnaces, which were called blowing-houses ; these
have all disappeared, but the name in some places still
remains. The tin ore as delivered at the smelting-house
contains from 60 to 75 per cent of white tin, the operations to extract which are conducted in reverberatory
furnaces.

A charge of about 3000 lbs. is mixed with culm sufficient
for deoxidation, and with lime or fluor spar to slag off the

silica. These must be thoroughly mixed with water
sufficient to prevent the draught from entangling the
fine tin. Immediately after charging, the door is closed
and a strong fire maintained until the whole mass runs
down, when the charge is well raked up, and the heat
raised to complete the smelting. This operation occupies
about 10 hours. The scoria, cooled by some damp smalls,
is then removed, the slags last raked off being reserved
for further treatment, because small shots of metal are
entangled in them. The white tin is then run into iron
pans and ladled into 3 cwt. moulds for liquation. The
blocks of tin are arranged in the same furnace, and a
moderate heat being applied, the tin melts slowly and
runs into the heated pan at the side of the furnace. As
the pile crumbles, so additional blocks are added, until
5 tons of tin have been collected. The residue is then
fused at a high temperature, and run into the other pan to
undergo another liquation; it contains nearly all the lead
and iron which contaminated the liquated blocks. When
the tin is tolerably pure this "sweating" process is
unnecessary, and is now seldom resorted to in Cornwall.
The tin, which has been kept molten by an auxillary fire,
is now refined by forcing bunches of green wood into the
bath; the ebullition that immediately follows produces
a drossy froth which is composed of stannic oxide, with
the oxides of the lead and iron that the liquation failed
to extract. After the removal of the wood, the bath is
allowed to settle an hour or so, during which it has
separated into zones of different quality, the purest
resting nearest the surface, and the second quality, or
common tin, occupying an intermediate position. The
tin remaining at the bottom contains about 5 per cent

of copper and lead, and has therefore, to be submitted to re-treatment. Refined tin contains about a quarter, and common tin nearly a half, per cent of metallic impurities. To smelt a ton of tin about 2 tons of coal are requisite. The loss in reducing the ore is often as high as a twentieth of the metal contained.

The uses of tin are manifold. It is largely employed in the tin-plate trade; is mixed with lead to form pewter; with copper to make bronze; and with copper and zinc to make brass and bell-metal. Tinfoil is made from it, and domestic utensils; and much is used by dyers in mordants, and by plumbers to compose solder. The importance of the tin-plate trade may be judged from the fact that 200 mills have been erected for their manufacture; but owing to the depressed state of the market for those goods, only 117 were in operation in 1882. During this year the aggregate production was 2,300,000 boxes weighing nearly 114,000 tons. Beside 12,189 tons obtained from the mines of Cornwall, and 50 tons from those of Devon, open quarries in the Roche district yielded 150, and streamworks 1655 tons of black tin; or a total of 14,045 tons for the two counties.

Tin mining in Cornwall has been injuriously affected by the discovery in Australia and Tasmania of vast deposits of alluvial tin, and the yield from them has so lowered the value of the metal, that few home mines can produce it at the ruling market price. In 1874 there were 230 mines in the two western counties returning tin, whilst in 1882 this number was reduced to a hundred. As the total weight of tin sold is about the same for both years, it follows that while the weaker companies have become defunct, the more important mines have greatly

increased their returns. The scant geological knowledge
we possess of the Australian tin fields scarcely permits the
formation of a definite opinion as to their continual
productiveness, though several papers read at the
Mining Institute seem to indicate that the alluvial tin
has accumulated during the secular degradation of the
granite. If this be so, it is by no means probable that
many tin veins of value will be discovered, and though
some years will pass before the deposits will be exhausted,
Cornishmen may look with some confidence to a
prosperous future. Meanwhile, the present regretable
depression has stimulated the energies of " One and All"
to obtain more efficient machinery, to more careful
economy in all operations, and to a steady resolve to have
the grasping cupidity of landowners checked by a fair
system of mining laws.

COPPER.

The tin industry, originated by the Phœnicians,
continued through the time of the Roman Empire to the
present day; though the production of that metal
fluctuated greatly. Copper does not possess the same
historic interest as tin, because its ores lying concealed in
veins usually under the tin, would attract less attention ;
and consequently the copper ore trade is of comparatively
modern growth. Early in the 17th century, copper—
which had been quarried in the celebrated mines of
Anglesea, and raised from the Saxon lodes as early as the
tenth century—began to be extracted from the Cornish

deposits.   At this time the ores of copper were so little
understood by the tin miners, that black and yellow
ores were thrown away as "deads," with the mundic.
In the middle of the 18th century, some of the Gwennap lodes yielded very rich work, *e.g* in 1757, copper
ore, to the value of £15,000 was raised in five weeks, from
Wheal Virgin, in Gwennap, at an expense of £300 only !
The first sales of which any record remains, took place
in 1726, when 2216 tons of copper ores appear to have
been sold.   In 1856 the amount raised had increased to
206,177 tons of six and nine-sixteenths produce, the
highest quantity ever produced.   The per centage of the
ores, which during the early decades of copper mining
was as high as 12, became reduced in 1850 to $7\frac{1}{2}$; and
during the last thirty years has varied from $6\frac{1}{8}$ to $7\frac{1}{8}$.
Fron 1726 to 1855, inclusive, the ore produced in
Cornwall and Devon realised over £50,000,000.

From 1855, owing to the exhaustion of the shallow
deposits, and the transition of the lodes from copper to
tin-bearing, there has been a gradual, but constant
decrease in the production ; so that in 1882, the copper
ore sold amounted to only 26,641 tons, which contained
an average of $6\frac{1}{4}$ per cent of pure metal.   With so small
an output, Cornwall has lost its reputation as a copper
producing county.

Some of the adits, ramifying through the mining
districts to drain off the surface water which percolates
through the rocks, are of great length, and when driven
through copper-bearing strata, the waters discharged
therefrom are more or less cupreous.   The most important one, running under the Parishes of Gwennap,
Kenwyn, St. Agnes, and Redruth, has—inclusive of its

branches—a total length of thirty-three miles.*   The
percolation of the pluvial waters through such a number
of veins, has collected in solution such a proportion of
the sulphates of copper and iron, that the former can be
profitably precipitated in " streaks " by the aid of scrap
iron.   Copper was first obtained by this method as early
as 1750, in St. Just; but cupreous waters rushed in
great volume from the Gwennap adit a century before
any means were adopted to check this waste (in 1824).
At Nangiles, where water retained so much sulphate in
solution that the pumps had to be lined with wood;
precipitating works were attempted in 1830, but as the
produce sold for only £9 per ton, the mode adopted to
gather the copper must have been very faulty.   Precipitating works now occupy numerous sites in the Carnon
Valley, and return copper dust worth 40 per cent.   The
waste of iron is considerable, as two tons of it are used
up to throw down a ton of copper.   The re-action by
which the metal is obtained, is, of course, well known,
the sulphuric acid of the cupric sulphate combining with
the metallic iron to form sulphate of iron, which is
carried away in solution.   Mr. W. J. Henwood estimated
that the water—1600 cubic feet per minute in 1850—
issuing from the Gwennap adit held in solution one part
in 600,000; if so, this quantity was sufficient to have
produced forty tons of copper annually, if all could have
been precipitated.   Since the stoppage of the Gwennap
mines, the cupreous value has been much lessened.

---

* The crosscut adit which unwaters the lodes of the Schemnitz
mines to a depth of about 300 fathoms, is eleven miles long.

COPPER ORE DRESSING.

As in tin mines, so in those of copper, a large proportion of the lode stuff raised is " deads," that is to say, the proportion of metal included being too small to be profitably separated, they are trammed away to the waste heap. As the ores are rarely so regularly disseminated throughout the matrix as to require fine pulverisation, the dressing operations are simple, and the machinery not nearly so complex and costly as those in tin mines.

The gangue from the cupreous portions of the lode receives small attention in the "stopes" where it is " broken," but usually reaches the dressing floors unsorted. The methods in vogue for raising the produce of the gangue have a generic similitude throughout the county, but they are varied to suit the different species of ore. The first operation—except in case of oxidised ores, the comparative quantity of which is insignificant—is to reduce the larger fragments of rock by passing them through a Blakes crusher, or by breaking them up by a sledge hammer, and rejecting the poorer veinstone ; this is termed "ragging." After this preliminary dressing, if the ore is moderately rich, a pile will be left fit for sale, and the rest goes through the Cornish crusher between the rolls of which it is broken down to the size of hazel nuts. Should there be no such machinery on the mine the "ragged " ore is " spalled," " cobbed," and " bucked " by the " bal maidens " to the requisite size. The ore is then "jigged" on an oblong sieve which is jerked up and down in a wooden case full of water, when the fines fall into the hutch below, the roughs remain on the sieve, and the stony matrix is scraped off them and thrown over the skimping, or waste heap. For sampling and selling, the picked ore from the ragging, must also be reduced to smelting or hazel nut size.

Should the copper ore be so poor as to demand fine crushing, or even stamping to separate its metallic contents, " buddling is rssorted to. But before washing, the ores are passed through a trommel in order to classify them into sizes, after which they are buddled, re-buddled and tyed, until the ore is brought up to a marketable state. By these various dressing operations, copper ores with less than two per cent of metal, can be raised to ten per cent. Some ore becomes so triturated in the stamps that it is floured, in which case slime separations and packing kieves become necessary. Although copper is not carried away with the slimes to such an extent as tin, there is still a great loss in mixed gangues, as so much ore remains in the rejected veinstone, and in the skimpings.

SAMPLING.

All the copper ores of Cornwall and Devon were formerly sold at weekly, but now at fortnightly ticketings at Redruth and Truro, and sent to the Swansea smelting works, where they are mixed with foreign ores. They are then submitted to a succession of calcinations, and fusions, until a good tough copper known as " best selected " is produced. Up to the close of the 18th century, copper ores were smelted at Copperhouse, near Hayle, but they are now shipped to Swansea, where cheaper fuel

and greater facilities for reduction exist. The *samplings*
which take place on the mines usually every month, are
conducted by men specially employed by the copper companies and are called "samplers." To ensure an average
sample, the ores are mixed by carrying them in hand-
barrows to raise rectangular piles about thirty inches
high; through the centre of which a trench is cut, and
from its sides a certain quantity of the ore is cut down
and mixed; and from the heap thus procured convenient
samples are impartially taken. The price per ton to be
given is then calculated from the result of dry assay which
is often two per cent below the true cupreous value.
From the price of the parcel 55/- per ton are deducted
for freight and cost of smelting.

HYDRO-METALLURGICAL PROCESSES.

The mining districts of Cornwall are bestrewed with
vast heaps of mining rubbish, which in many cases contain an aggregate of metallic value sufficient to leave a
profit after subjection to economical treatment by means
of a *wet process*. Many shallow portions of lodes—frequently above the adit—contain enormous reserves of
similar poor ores, which enclose arsenic, copper, silver,
and other metals; and these, though worthless in combination, can, in many places, be separated at a considerable
profit. In 1820 coppery mundic was roasted to sulphates,
and the soluble copper thrown down on scrap iron; and
later, copper was similarly precipitated from the cupreous
waters of the Gwennap adit. The first serious attempt
to treat mixed ores were made at Greenhill and Luckett,

near Callington, where about the year 1874 considerable works were installed to treat the ores proceeding from the pyritous lodes worked on the slopes surrounding the granite boss dominating the Tamar river between Callington and Tavistock. These ores exist in quantities almost inexhaustible, and are rich in arsenic and silver; are cupreous, and often stanniferous. At NewGreat Consols during 1874, and following years, the wet reduction of such ores was in successful operation, but through injudicious interference with the management on the part of the London officials, this promising attempt to utilise poor mixed ores collapsed. The same fate overtook the Cornwall Chemical Company, whose works at Greenhill, and other places, though well conceived and arranged, were erected on a scale for which the capital subscribed was utterly inadequate.

It may be admitted frankly that the changeable character of Cornish ores, and their siliceous gangue, which prevents the sale of residues to the iron makers, militates against the process; still the Spanish pyrites, whose great merit is in the constancy of its composition, contains only two ounces of silver to the ton, and has to be freighted to England. The value of the arsenic in the ores of the Callington mines may be considered to counterbalance the disadvantages acknowledged. It is unquestionable that—as the mining and treatment of the arseno-argentiferous ores would exceed little, if at all, a cost of twenty shillings a ton—wet process reduction intelligently and honestly conducted, would open an excellent channel for the investment of capital, and for employment of numbers of the non-employed mining population. The process of reduction adopted at New

Great Consols was simple, and involved no complicated chemical reactions. The crushed ores were introduced into calciners, and the sublimed arsenic collected in long flues and refined; the ores were then roasted with salt and lixiviated, after which the chlorides of copper and silver were precipitated together in a tank containing scrap iron; and finally, the residues were dressed in the usual manner to separate the tin oxide. This process would have been much improved if the silver had been separated from the copper.

The use of copper, pure or alloyed, is so universal that it would occupy an inconvenient amount of space to enumerate them all. It is employed extensively for bolts, sheating, and nails, for vessels; for wire, silver, tubes, coins, and domestic utensils; and for bearings, pipes, and boilers, in factories, distilleries, and steamships.

LEAD.

Possibly the Phœnicians obtained from the Damnonians lead as well as tin, and the Latins, who wrought lead mines in Derby and Cardigan, may have encouraged its production in the western counties. Lead mining in Cornwall has never attained to any great importance, and has been somewhat spasmodic, owing to the sporadic discoveries, both in space and time, of rich deposits, and the celerity with which galena can be extracted. Lead is often associated with limestone, but in Cornwall the connection does not exist, for although some mines have been worked in the slightly calciferous Devonian slates in

the vicinage of Padstow, the most important lodes occur
in clay-slates, which are distinctly subordinate to the
granite, and are productive indifferently whatever their
direction.   Thus the Trelawny great lode had a north-
north-east bearing, whilst the celebrated West Chiverton
lode bore south-west.

Penrose Lead Mine, near Helston, which was working
in the 17th century, and Old Garras in 1720, and again
about the year 1820, yielded galena, containing a hundred
ounces of silver to the ton of lead.   In the present
century the most prominent mines have been Shepherds,
Wheal Golding, Penhale, East Wheal Rose, Trelawny,
and West Chiverton.   In 1845 Cornwall produced 6,063
tons of lead, but in 1853 the total had fallen below 5,000,
from 1867 to 1871 the large sales of galena at West
Chiverton brought the total up to about 6,500 ; since
then the returns have generally fallen off, until in
1882 the sales amounted to only 454 tons of lead
worth £10 per ton.   From 1853 to 1882 inclusive,
the 118,000 tons of lead produced from the galena sold,
contained an average of forty-one ounces to the ton, or
a total of nearly five million ounces of silver.

The lead ore raised has been always smelted in the
county, but some of the works have gone to ruin.   At
Par and Devoran, until lately, lead ore was reduced and
the metal desilverised.

The smelting of lead sulphide is not difficult, and
if free from impurity, it could be reduced without the
aid of any flux by roasting and fusion in one furnace
operation, with an addition, towards the end of the
melting, of some carbon to reduce the oxy-sulphide
formed.   When the lead ore is very poor or fouled by

mundic, &c., a preliminary roasting is advisable. About thirty hundred weight is smelted at a charge in reverberatory furnaces, the time occupied varying according to the *modus .operandi*, from six to ten hours. The loss of lead in reduciug the sulphide is sometimes as much as seven per cent., but nearly three are recovered from the slag and fume chambers. The consumption of fuel required to smelt a ton is about 15 cwt.

Lead is applied in the manufacturies to the making of paints, tanks, pipes, and domestic utensils, &c., and for roof and acid chambers. The following tabular statement gives the yield of tin, copper, and lead, obtained from the mines of Devon and Cornwall, as far back as the records reach :—

Statistics of Tin, Copper, and Lead in Tons.

	TIN.	COPPER ORE.	LEAD.
	Devon &Cornwall.	*Cornwall.*	*Cornwall.*
	WHITE TIN.		
*1726 to 1735		64,800	
1736 to 1745		75,520	
1746 to 1755	? 26,160	98,790	
1756 to 1765	26,690	169,699	
1766 to 1775	28,358	264,273	
1776 to 1785	26,837	304,133	
1786 to 1795	33,825	? 385,000	
1796 to 1805	28,109	564,037	
1806 to 1815	24,758	726,308	
1816 to 1825	† 36,195	926,271	
1826 to 1835	42,510	1,352,313	
1836 to 1845	? 73,587	1,486,840	

* From Charles I. to 1750, the returns of white tin averaged about 1500 tons per annum.

† To 1825 the returns are for Cornwall only.

	TIN.	COPPER ORE.	LEAD.
	Devon & Cornwall.	*Cornwall.*	*Cornwall.*
	BLACK TIN.		
1846 to 1855	95,673	1,622,152	
1856 to 1865	118,588	1,448,833	47,753
1866 to 1875	139,685	615,966	47,241
1876	13,688	43,016	2070
1877	14,142	39,225	2167
1878	15,045	36,871	1022
1879	14,665	30,371	545
1880	13,737	26,737	570
1881	12,898	24,510	409
1882	14,170	25,641	454

SILVER.

Besides the association of silver with galena, which is so pronouced in the lead lodes of Cornwall, it is found native, and in a mineralised state in every mining district. The first discovery of argentiferous ores is said to be made at Huel Mexico, north of Truro, where it occurred in nests of chloride, and native. Early in the present century much silver was obtained from Huel Brothers, Dolcoath, and Herland Mines ; and later at West Darlington, Ludcott, Carnbrea, and Huel Duchy, near Callington ; in the latter mine a shallow course of ore enclosed in slate near the granite, was of extreme richness. Although Carn Entral and Druids lodes yielded much silver, it is usually found in cross veins as at Herland Mine, where vitreous silver ore and the metal itself were dispersed through the veins, with quartz and mundic,

Pubd by H.Besley, Directory Office   *Compact Felspar and Hornblendic Rocks.*

GURNARDS HEAD, CORNWALL

South Street, Exeter

which, near the surface had been weathered to gossan.

The lodes around the Gunnislake granite often include silver, which, in many instances would pay for working. Green and brown chlorides of silver were found on the back of Trelawny lead lode, and in the gossans of the Dolcoath lodes west of Camborne. In 1858 North Dolcoath sold eight tons of chloride ores for more than £100 per ton, and in the year following, some was sold for more than £500 per ton. The occurrence of silver in the gossans of the copper lodes is not unusual, and much has been thrown over the burrows in ignorance of its value : it is not improbable that a careful examination of the backs of gossans would be requited.

ARSENIC.

This metal in the form of arsenical pyrites is the close companion of tin ore, from which it is separated—after concentration into " witts " by calcination. This roasting changes it into an oxide that sublimes, and being conducted through long flues, it falls in an impure condition to the bottom, whence at intervals it is withdrawn to be refined in works specially erected at Bissoe Bridge, Roseworthy, and Hayle. With the exception of East Pool, West Seton, and South Crofty few mines in the West sell crude arsenic. In the Gunnislake district the mines of New Holmbush and Okel Tor raised in 1882 no less than 10,058 tons of arsenical pyrites, which were sold for twenty shillings a ton and produced 2258 tons of refined arsenic. The largest works in the two counties is at

N

186 Geology of Cornwall.

Devon Great Consols, where nearly 3000 tons of refined arsenic are produced annually. From Cornwall 3473 tons, and from Tavistock 3996 tons of arsenic were obtained in 1882, and sold for £6 : 11 : 7¼ per ton. No mines in the kingdom produce arsenic, outside the limits of the Stannaries.

IRON.

Iron is widely distributed in the veins of all the metalliferous tracts, either in the form of sulphide in the depths of the mine, or as oxide in the gossans of lodes and crosscourses, and as carbonate in the Perran lode ; many of the rarer combinations are frequently found in beautiful crystals.

The most noteworthy deposits of iron are subordinate to the Hensbarrow granite. They occur on the backs of some of the east and west lodes that traverse the killas near its junction with the granite; and in the crosscourses, the quantities raised from which, at Restormel, Coldvreath, Pawton, and Tolbenny, have been enormous. Ferruginous masses of less magnitude are found in the Hensbarrow and St. Just granites. The hematites of Cornwall are much esteemed in Wales as a flux for other ores, but owing to the expense of conveyance, they can only be sent there when high prices rule. Should it ever be required, the county could furnish immense supplies. The ores produced during the year 1882 were mostly brown hematite, and the tonnage was as follows :

Coldvreath	...	550
Restormel	...	848
Perran	...	4351

Equal to 5749 tons.

The produce of the ores was $46\frac{1}{4}$ per cent, and the average price half a guinea per ton.   Probably *Restormel Lode* has yielded more iron than any other in the county. It enclosed beside hydrous oxide, some hematite, specular ore, and a little manganese, with many other associated minerals in minute quantity.

The **Perran Iron Lode** from its persistent continuity, its great width, and the mass of mineral matter which fills it, is held to be the most remarkable vein in the west.   It courses up the cliffs at Ligger Bay in a ruddy mass a chain wide, and extends in a south-east-by-east direction through Gravel Hill, Duchy Peru and Deer Park, to Mitchell, a distance of six miles.   The ores both limonite and chalybite have been raised in a fitful way for a number of years, but it was not until the iron fever of 1870 that the mines were brought into prominent notice by Mr. Roebuck, who promoted the Cornwall Minerals railway company, whose line now winds amongst the Perran mines.   Much of the white iron, or spathose ore, was at one time sent to Dowlais for making spiegeleisen, for which its purity and perfect exemption from phosphorus, eminently fits it.

The back of the lode has been oxidised to brown hematite to a depth of from fifty to a hundred feet, but below the influence of the atmosphere, a solid and practically exhaustless dyke of ferric carbonate seems to extend the whole length of the lode.   This spathose iron is rendered the more valuable, on account of the reported large proportion of manganese, which amounting to seven per cent (?) is equal to the most manganiferous carbonates in England ; it is also said to contain several ounces of silver to the ton.   The lode at times reaches a width of 120 feet, and its dip—discordant with the strata—is about 35° toward the south-west. Much of the space between the walls is occupied by quartzose and brecciated matter, so that the width of workable ore would not perhaps average over fifty feet. It is much to be deplored that such a magnificent lode, so well opened out and provided with efficient machinery, and in direct railway communication with a not distant port, should be unable to longer struggle against a depressed and glutted market.

MANGANESE.

Minerals having manganese for their base, are not in Cornwall associated with hematites to any great extent. In the carboniferous strata, whose basset edges, intermixed with greenstone, contour from Petherwyn to Lidford, many small deposits have been worked. Pyrolusite was raised in the Tregoss Moor, near Roche, in 1754 for use in glass making, but was not discovered in the Launceston district until 1815. The usual selling price has been about £3 per ton ; in 1879 it dropped to 36/-, but in 1882 its value has reached as high as £3 : 10 : 0 per ton.

The yearly production of manganese in Devon and of iron ores in Cornwall is tabulated below.

Minerals produced—in Tons.

	PYRITES. Cornwall.	HEMATITES. Cornwall.	MANGANESE Devon.
1854	128		
1855—1864	129,056	284,206	4,988 ?
1865—1874	52,575	204,130	36,413
1875	7,223	11,403	3,205
1876	8,244	18,390	2,705
1877	14,290	4,963	2,496
1878	3,203	1,308	1,404
1879	1,049	400	607
1880	6,369	15,865	2,383
1881	14,910	7,460	1,845
1882	11,343	5,749	862

GOLD.

This most coveted metal has been taken from all the tin stream works in Devon and Cornwall. No doubt in very early ages it was plentifully scattered through some of the tin layers, as an old record relates an anecdote of some streamers, who, in 1753 brought to the Blowing House a parcel of tin with so mnch gold in it, that the smelter jumped to the conclusion that it must be mundic, and rated the tinner soundly for bringing him imperfectly washed tin. Gold has been drawn most largely from the stream-works at Carnon, Pentuan, and Ladock, but in modern times the amount obtained has been very insignificant. The gold nuggets and spangles were detached from the rocks by denudation and gradually rolled down the stream.

Minute particles of the precious metal have also been met with in veins, and in the mundic of some lodes. Gold occurs in the mines of Sperris, Sparnon, Garras, Carn Brecon, Sheepstor, and others.

ZINC.

Zinc-blende, the "black jack" of the Cornish miner, is plentifully distributed in some lodes, and few are destitute of it. In Pryce's time the blende from copper lodes was used to make brass. The sulphide of zinc frequently occurs in large bunches, and is often associated with iron and copper pyrites, and with galena, and sometimes with cassiterite. The ores now sold proceed almost entirely from the Duchy Peru lode, which in 1882 yielded 4059 tons of blende, that averaged a produce of 48 %, and realised £3 per ton. Mellanear, Violet Seton, and West Chiverton sell only unimportant quantities, though a few years since the latter mine raised thousands of tons yearly.

Calamine has been remarked, though small in quantity. Zinc is rolled into sheets for roofing and lining cisterns, into wire and piping; and is used to make brass, to galvanise iron, and to produce electricity.

RARER METALLIC ORES.

Wolfram —The tungstate of iron and manganese is found in all the mining districts, and prevails in the mines along the northern margin of the Carn Menellis granite, most markedly at East Pool, which mine, in 1882, sold 58 tons of wolfram at £13. It contaminates the tin ore in some of the mines in the Hensbarrow moors, and was mixed in larger proportion with the ores of Drakewalls mine, near Calstock. Tungstic acid and tungstate of soda are used in the manufactures

Uranium.—The ores of this metal have occasionally been met in mines situated in the parishes of Saint Just, Gwinear, Camborne, Illogan, Redruth, St. Agnes, St. Stephens, and St. Austell. In Huel Providence and Trenwith it was mingled with the copper ore in quantity sufficient to deterioate its value. Small parcels of pech-blende (oxide of uranium) were a few years since collected at Wheal Owles and East Pool, and sold at prices oscillating between £100 and £200 per ton. Lately some lime-uranite has been found at South Terras, and sold at the rate of £500 per ton. Uranium is used in the ceramic arts for colouring.

Bismuth.—Bismuth, native and mineralised, has been met in Saint Just and Saint Ives ; also in Gwinear, Illogan, Redruth, and Saint Austell. It occurs sporadi-cally in small pieces in the lodes of Restormel and Fowey Consols, in some mines near Calstock, and at Ivey Tor. It is not often that it exists in quantity, notwithstanding that in 1876 and a few previous years, Wheal Owles and East Pool sold a few hundred weight. Alloyed with lead and tin, it is used for stereotyping, and for preparing pewter ; and with mercury it is used for silvering globes, &c.

Nickel and Cobalt.—Although nickel is found in some abundance in Germany in the mineral called kup-fernickel, its occurrence in Cornwall is uncommon. The nickel is generally mixed with cobalt, and sometimes with bismuth. It has been raised from Botallack, Dolcoath, East Pool, Huel Sparnon, Trugoe and Drakewalls, and

especially at St. Austell Consols, which sold during the
decade following the year 1860 considerable quantities of
low quality ores. A few years since East Pool and
Botallack sold these ores in insignificant quantity. Nickel
is used in the arts for cutlery and domestic utensils, and
cobalt for ensuring permanency of colours to china, glass-
ware and metal.

Molybdenum.—This metal in the state of sulphide
is rarely met, but is sometimes mingled with tin ores. It
has been noticed in St. Ives, Lelant, Gwinear, Gwennap,
and Calstock. Pure molybdenite is said to have a high
value.

Antimony.—None of the ores of this metal are now
raised in the county, but the grey sulphide occurs in slates
associated with trappean rocks, in the vicinity of Padstow,
St. Germains and St. Austell. It was worked at St.
Stephens in 1758, and in 1778 there were antimony
works at Point, near Devoran.

Minerals produced in Cornwall—in Tons.

	Zinc Blende.	Arsenic.	Tungsten. W.
1854	638	477	
1855 to 1864	24,422	5,801	78
1865 to 1874	11,981	17,902	231
1875	3,087	2,412	46
1876	4,414	2,557	23
1877	4,991	1,718	15
1878	4,483	1,843	10
1879	3,202	1,655	13
1880	4,440	1,356	1
1881	7,793	2,775	54
1882	4,608	3,473	58

CHAPTER VII.

QUARRIES.

Both the crystalline and schistose rocks of Cornwall
yield building material of excellent quality and durability,
and that they are much sought after is made patent by
the extent of the exportation. The **granites** from the
quarries of Lamorna near Penzance, of Mabe close to
Penryn, of Luxulyan near St. Austell, and of the Cheesewring north of Liskeard, have long been celebrated for
their admirably fine grain and beautifully grey colour.
As the durability and, as a whole, the appearance of
granite must be attributed to the felspar, the condition of
this important ingredient is always minutely examined
before employing it in building, because the felspar though
compact often contains the hidden germ of decomposition.
Some of the excavations are very large, that of the
Cheesewring being more than 100 feet deep and 400 feet
wide. The granite can be cut out in large masses, and
in the porphyritic granite of Luxulyan monoliths of 20
feet or more can be cleaved. The granite is squared or
roughly blocked, and is exported to most of the towns
in the south and west of England, Liverpool and London
taking large quantities.

No statistics of the present production are available,
but in 1858 the number of tons produced appears to have

been about 80,000, and its value at the quarry about 20/-
per ton.

Elvan. This rock, procured from the courses of porphyry which traverse most of the county, is quarried for
rough building stone and for road metal. Some elvans
when first quarried are soft, but the exposure of a few
days, dries and hardens them ; one of these at Wheal
Prudence, in St. Agnes, is much sought after to make
troughs. The Pentuan elvan has been long known for
its beautiful grain and colour. The St. Wenn elvan is
very handsome when polished.

Serpentine. Some twenty years since, the quarries
on the Lizard were much worked for this rock, which
possesses a variety of agreeable shades of green and red
with a silky lustre. Large buildings fitted with appropriate machinery, were put up at Penzance to fashion it
into ornaments, monoliths, &c. ; but the manufactory did
not prosper and the trade has dwindled, but there are some
quarries at work near Cadgwith, and a mill at Poltesca
for the turning of ornaments, &c.

Roofing Slate. In the Upper Devonian slates
along the coast west of Camelford, are several quarries
producing this material. The principal, and indeed the
only one largely worked, is that known as old Delabole
quarry, which has supplied roofing and flooring slates for
the past 300 years. Steam power was introduced in
1837, and the output increased to such an extent, that in
1847 five steam engines and 1000 persons were employed.
In 1882 the workmen numbered 350, and the quantity of
manufactured slates, slabs, &c. equalled about 10,000 tons.
The lightness, strength, and durability of the Delabole
brand is so highly appreciated, that the slates find markets

on the Continent, in the West Indies, and in America.
The total quantity produced by the Cornish slate quarries
during the above year was 11,680 tons, said to be worth
40/- per ton. The quarry is about 400 feet deep.

Hornblendic Rocks. Greenstone is much used for
macadam all over the county, and hornblende schist is
extensively quarried at Porthalla near the mouth of the
Helford river for re-metalling the highways.

Clay Slates. These are quarried everywhere, and
many of them yield a most durable and sightly building
stone.

Limestone. Few quarries have been opened in good
limestone in Cornwall, because of its scarcity, but vast
quarries are worked in the pure crystalline lime rock of
Plymouth. The blue limestone beds of Towan Head
near Newquay are quarried for hydraulic cement.

CHAPTER VIII.

MINE WATER.

The enormous expense of draining Cornish mines, especially when they are deep, is the primary reason why so few of the tin mines can pay dividends. To increase the efficiency of the draught engine and to reduce the cost of fuel, the unsparing energies of generations of engineers have been employed; until, as a pumping machine, the Cornish engine has attained such perfection, that it is in demand throughout all mining countries.

To attrap the surface water, and to hinder its descent into the depths of the mine, long and expensive tunnels have been driven into the hills at the lowest level attainable; and at the surface, hollows are filled and channels often made, to accelerate the discharge of the rainfall. It has been calculated that half the water discharged by the Gwennap Great Adit, was raised from an average depth of 190 fathoms, and that £20,000 a year were saved in fuel by its existence. The temperature of the adit water, which was 60½° in winter, and 68° in summer, was more than 12° above the mean temperature even in summer.

Rains flow into the levels at various intervals after their fall on the surface, in some mines the water percolates quickly, in others two or three months are

required, and even longer. The water permeates more
freely, through granite than slate, though in the former
it does not penetrate in such large quantities to the
deeper portions of the mine ; massive crystalline slate
includes less water than the schists. The numerous beds,
joints, veins, and crosscourses in the clayslate, afford
great opportunities for aqueous circulation throughout
large areas; so that when the nature of the ground and
the position of flucans are imperfectly known, it is not easy
to recognise the area which a shaft would have to
unwater.

The water pouring from the metalliferous granite is
considered potable by the miners, but on the schistose
rocks, salts of the prevailing metals are often held in
solution, and in the mines near the sea the water is more
or less salt.

TEMPERATURE OF MINES.

The late Mr. W. J. Henwood, F.R.S., who during
half a lifetime devoted untiring energy to the study of
the geological phenomena connected with mining, has
presented us with a series of very valuable tables
illustrating the temperature of the rocks and lodes underground. The general results which he has recorded
are doubtless approximatively correct, but is well to
notice that owing to changes introduced by the excavation
of shafts and levels—which introduce air and water from
the surface—the tables must necessarily register a lower
degree of heat than would be normally the case. The

following are the temperatures given by Mr. Henwood
as a mean of all the Cornish districts :

Depth 30 fathoms, temperature 55°
„   72   „   „   61°
„   127   „   „   67½°
„   173   „   „   78°·
·, 240   „   „   85½°

The heat of the schistose rocks is considerably higher
than that of the granite, and this temperature has a
greater proportionate value as depth is gained ; so that
though at 30 fathoms the heat of slate is only 55·9° as
against that of granite 52·7°, yet at 240 fathoms it is
89·4° to 76·15° F. Thus in the granite it requires 51 feet
of depth to gain a degree Fahrenheit of heat, whilst in
the slates 37 feet appears to be sufficient. Naturally the
rocks are hotter than veins, the heat increasing from the
walls inwards, and lodes would possess a higher
temperature than crossveins, because of the water
running through the latter, and on account of the absence
of workmen. Mr. Henwood convinced himself that tin
lodes are colder than tin and copper lodes, and that those of
copper are the warmest ; and that although the difference
is not great, yet it is sensible to the miners. In some
levels, owing to hot currents of water welling up in them,
the heat has been almost unendurable. At the 320
fathom level in Tresavean it was over 90°, and in the
deep level of the United mines as high as 108° F.

Before the opening of the Great Sutro tunnel, the water
issuing from the Comstock lode, at a depth of 2700 feet,
had a temperature of 157° F, and men falling into it
have been fatally scalded.

TABLE OF MEAN TEMPERATURES AT SURFACE.

	Penzance, 1836.	Truro, 1883.
Mean of year	54°F	50·8°F
Hottest month	65°	60·9
Coldest month	43°	39·7
Rainfall in inches	44·70	40·73
No. of wet days	178	195
No. of dry days	187	170

INDEX.

o

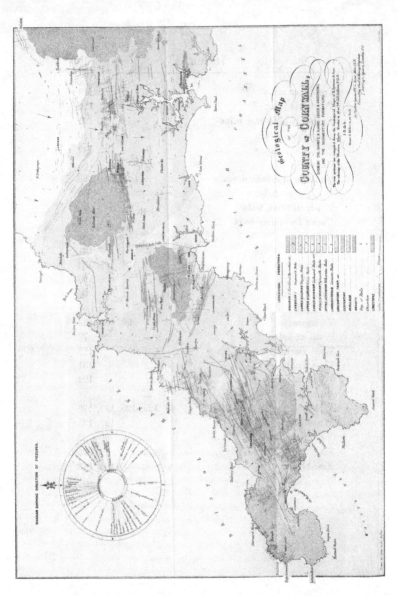

The material originally positioned here is too large for reproduction in this reissue. A PDF can be downloaded from the web address given on page iv of this book, by clicking on 'Resources Available'.

Printed in the United States
By Bookmasters